Aymen Gatri

Next Generation Optical Wireless Communication Systems

A Systems Approach

disserta
Verlag

**Gatri, Aymen: Next Generation Optical Wireless Communication Systems.
A Systems Approach**, Hamburg, disserta Verlag, 2019

Buch-ISBN: 978-3-95935-499-8
PDF-eBook-ISBN: 978-3-95935-500-1
Druck/Herstellung: disserta Verlag, Hamburg, 2019
Covermotiv: Pixabay.com

Bibliografische Information der Deutschen Nationalbibliothek:
Die Deutsche Nationalbibliothek verzeichnet diese Publikation in der Deutschen Nationalbibliografie; detaillierte bibliografische Daten sind im Internet über http://dnb.d-nb.de abrufbar.

The Book is based on the Dissertation at the Northumbria University Newcastle, the thesis title is:"Performance optimization mechanisms for optical wireless communication systems".

Das Werk einschließlich aller seiner Teile ist urheberrechtlich geschützt. Jede Verwertung außerhalb der Grenzen des Urheberrechtsgesetzes ist ohne Zustimmung des Verlages unzulässig und strafbar. Dies gilt insbesondere für Vervielfältigungen, Übersetzungen, Mikroverfilmungen und die Einspeicherung und Bearbeitung in elektronischen Systemen.

Die Wiedergabe von Gebrauchsnamen, Handelsnamen, Warenbezeichnungen usw. in diesem Werk berechtigt auch ohne besondere Kennzeichnung nicht zu der Annahme, dass solche Namen im Sinne der Warenzeichen- und Markenschutz-Gesetzgebung als frei zu betrachten wären und daher von jedermann benutzt werden dürften.

Die Informationen in diesem Werk wurden mit Sorgfalt erarbeitet. Dennoch können Fehler nicht vollständig ausgeschlossen werden und die Bedey Media GmbH, die Autoren oder Übersetzer übernehmen keine juristische Verantwortung oder irgendeine Haftung für evtl. verbliebene fehlerhafte Angaben und deren Folgen.

Alle Rechte vorbehalten

© disserta Verlag, Imprint der Bedey Media GmbH
Hermannstal 119k, 22119 Hamburg
http://www.disserta-verlag.de, Hamburg 2019
Printed in Germany

ABSTRACT

In this day and age, free space optical (FSO) enable the deployment of a new category of products that can transmit voice, data, and video at bandwidths up to 2.5 Gbps at distances up to 4 km, over any protocol. This optical connectivity doesn't require expensive fibre-optic cable or spectrum licenses. FSO is reliable due to the capability of FSO systems to provide truly broadband and secure communications, as well as their immunity to interference from other sources of optical radiation. The principal challenge facing FSO technology is to achieve 100% link availability in all weather conditions. While rain, fog, haze, turbulence and aerosols all attenuate the optical signal to a certain level, fog is considered to be the main impairment in FSO systems. Thick fog resulting in over 300 dB/km of signal attenuation can reduce the transmission span from a few kilometres to just 100 m or so. Turbulence (i.e. scintillation) also results in reducing fade margins from 4 to 10 dB for FSO links of 1 km length or less, which is well below the margins for atmospheric attenuation. In the real-world environment, it is very challenging to measure the effect of atmospheric fog under diverse circumstances. This is due to several reasons: (i) the longer observation time required and the lower probability of reoccurrence of dense fog events for visibility $V < 0.5$ km, and (ii) the difficulty in controlling and characterising aerosols in the atmosphere, due to the inhomogeneous presence of aerosols along the FSO link path. This book examines and analyses the performance of a terrestrial FSO system by investigating the impact of a number of modulation techniques on mitigating the atmospheric impairments. A dedicated indoor atmospheric chamber is designed to carry out tests and measurements in a controlled manner and mimic the real outdoor atmosphere. The experimental results are compared with predicted data for the range of modulation techniques tested in the presence of atmospheric turbulence and fog, including binary phase shift keying subcarrier intensity modulation (BPSK-SIM), 2-pulse position modulation (2-PPM), 4-pulse position modulation (4-PPM) and hybrid pulse position modulation binary phase shift keying subcarrier intensity modulation (BPSK-SIM-PPM).The results show that BPSK-SIM-PPM offers a similar performance to 2-PPM, a superior performance to BPSK-SIM, while having the same bandwidth, and an inferior performance to 4-PPM under turbulent conditions. Furthermore, the experimental investigation indicates that 4-PPM is more resistant to turbulence compared to BPSK-SIM. The improvement of the link performance by optimising the beam spot size using combinations of mirrors is also investigated. In addition, the effects of low to high visibility on the FSO link BER performance in the presence of fog are measured and investigated. The obtained results indicate the dependency of the performance of the FSO link on the fog intensity variation. Moreover, the experimental results show that the impact of severe fog induced attenuation is greater on the receiver than the transmitter. Finally, the effect of fog on an FSO system employing quadrature phase-shift keying (QPSK) and operating at various carrier wavelengths is also studied.

TABLE OF CONTENTS

List of Figures v
List of Tables viii
Glossary of Abbreviations ix
Glossary of symbols xiii
Dedications xx

1. Introduction 1
    1.1. Background 1
    1.2. Research Motivation and Justification 4
    1.3. Research Objectives 8
    1.4. Book Organisation 10
    1.5. Original Contributions 12

2. Fundamentals of FSO 15
    2.1. Introduction 15
    2.2. Overview of FSO Technology 15
    2.3. Features of FSO 16
    2.4. FSO Applications 19
    2.5. The FSO System 20
        2.5.1. The Transmitter 21
        2.5.2. The Atmospheric Channel 23
        2.5.3. Atmospheric Attenuation 24
        2.5.4. The Receiver 27
    2.6. Noise in Optical Detection 28
        2.6.1. Thermal Noise 29
        2.6.2. Photon Fluctuation Noise 29
        2.6.3. Dark Current Shot Noise 30
        2.6.4. Background Radiation 30
    2.7. Eye Safety and Standards 31
    2.8. Summary 34

3. The Atmospheric Turbulence Model 35
    3.1. Introduction 35
    3.2. Optical Turbulence 36

3.3. The Lognormal Turbulence Model 39
3.4. Spatial Coherence in Weak Turbulence 45
3.5. Summary 46

4. Fso Modulation Techniques 47
4.1. Introduction 47
4.2. On-Off Keying 49
    4.2.1. OOK-NRZ 49
4.3. Pulse Position Modulation 51
4.4. Subcarrier Intensity Modulation 54
4.5. Quadrature Phase Shift Keying (QPSK) 57
4.6. Summary 60

5. Design of the Atmospheric Channel 61
5.1. Introduction 61
5.2. The Atmospheric Chamber 63
5.3. The Intelligent Weather Station 65
5.4. The Controlled Turbulence Simulation Channel 67
5.5. The Controlled Fog Simulation Chamber 71
5.6. Summary 75

6. Performance of FSO links under Controlled Atmospheric Turbulence 77
6.1. Introduction 77
6.2. BPSK-SIM under Controlled Turbulence 78
6.3. Hybrid BPSK-SIM- PPM FSO with Turbulence 88
    6.3.1. Introduction 88
6.4. Results and Discussion 90
    6.4.1. Simulation Results 90
    6.4.2. Experimental Results 94
6.5. Summary 97

7. Performance of FSO under CONTROLLED FOG Conditions 99
7.1. Introduction 99
7.2. Coherent Detection of BPSK FSO Links with Fog 100
    7.2.1. The Characterization of Fog Attenuation Uisng a Laboratory Chamber 100
    7.2.2. Experimental Setup 102
7.3. Performance Analysis of QPSK FSO at Different Wavelengths under Fog Conditions 109

      7.3.1. Mathematical Model 110

           7.3.1.1.   Link Budget 110

           7.3.1.2.   Atmospheric Attenuation Modelling 111

      7.3.2. QPSK FSO with Fog 112

  7.4. Performance Analysis of Hybrid PPM-BPSK- SIM, PPM and BPSK-SIM FSO with Fog 118

      7.4.1. Characterisation of Fog-induced Attenuation Using A Laboratory Chamber 119

      7.4.2. Hybrid 4-PPM-BPSK-SIM, BPSK-SIM and 4-PPM with Fog 122

  7.5. Summary 127

8.   **Conclusions and Future Works** **129**

9.   **References** **134**

LIST OF FIGURES

Figure 1-1: A schematic illustrating the research roadmap and the original contributions. 14

Figure 2-1: The overview of the EM spectrum with its nominated frequency bands [66]. 17

Figure 2-2: Block diagram of a terrestrial FSO system. 20

Figure 2-3: Comparison between Rayleigh, Mie scattering and Mie Scattering with larger particles [49]. 25

Figure 2-4: Response/absorption of the human eye at various wavelengths [53]. 32

Figure 3-1: An atmospheric channel with velocity fluctuations and turbulent eddies. 37

Figure 3-2: Normalised log-normal pdf for a range of irradiance variance σ_I^2 values. 44

Figure 4-1: The power spectrum of the transmitted signals for OOK-NRZ and RZ [62]. 51

Figure 4-2: Time domain waveforms for 4-bit OOK and 16-PPM. 53

Figure 4-3: BER performance for OOK (NRZ and RZ) and 4-PPM [49]. 54

Figure 4-4: Block diagram of a BPSK subcarrier intensity-modulated FSO link. 56

Figure 4-5: QPSK symbol constellation with Gray coding. 59

Figure 4-6: Heterodyne T_x block system [148]. 60

Figure 5-1: A snapshot of an atmospheric chamber in the lab. 64

Figure 5-2: IWS signal-flow block diagram. 66

Figure 5-3: A snapshot of the Conrad Digital Radio Weather Station WS1600 [215]. 67

Figure 5-4: A block diagram of an FSO experimental system with turbulence simulation. 67

Figure 5-5: The laboratory turbulence chamber activated by hot and cool air. 69

Figure 5-6: The laboratory controlled fog and FSO link setup. 71

Figure 5-7: ROF scenario considering an FSO link length of 1 km. 72

Figure 5-8: Comparison between the measured mean ROF attenuation and the mean lab generated fog attenuation. 72

Figure 5-9: Block diagram of the experimental setup used to measure V over the length of the FSO link. 73

Figure 5-10: A block diagram for the measurement of fog induced attenuation and link visibility. 75

Figure 6-1: The laboratory turbulence chamber which is excited by hot and cool air. 79

Figure 6-2: The chamber and FSO link setup in the laboratory. 79

Figure 6-3: Sketch of diverted beams due to the turbulence source being positioned near the transmitter. 81

Figure 6-4: Sketch of diverted beams due to the turbulence source being positioned near the receiver. 81

Figure 6-5: The block diagram of a subcarrier BPSK system with two subcarriers. 83

Figure 6-6: Histogram of the received signal in cases of: (a) no turbulence, (b) turbulence in the middle of the chamber, and (c) turbulence near the transmitter, for weak turbulence ($\sigma_I^2 = 0.12$). 86

Figure 6-7: The measured eye diagram of the received BPSK signal in cases of (a) no turbulence and a data rate of 50 Mbit/s, (b) weak, middle of the chamber turbulence with $\sigma_I^2 = 0.1$ and a data rate of 50 Mbit/s, and (c) near receiver turbulence and near transmitter turbulence with $\sigma_I^2 = 0.09$. 87

Figure 6-8: Block diagram of an L-PPM-BPSK-SIM FSO system showing the building blocks of the (a) transmitter, and (b) receiver [169]. 89

Figure 6-9: Simulated and predicted BER against SNR for 2-PPM with and without turbulence for a range of Rytov variance values. 91

Figure 6-10: Simulated and predicted BER against SNR for BPSK-SIM with and without turbulence for a range of Rytov variance values. 92

Figure 6-11: Simulated BER versus SNR for 4-PPM, BPSK and PPM-BPSK with turbulence for a Rytov variance of $\sigma_I^2 = 0.1$ 93

Figure 6-12: Simulated BER versus normalised SNR for 4-PPM, BPSK and PPM-BPSK with turbulence for a Rytov variance of $\sigma_I^2 = 0.3$. 93

Figure 6-13: Simulated BER of 4-PPM, BPSK and PPM-BPSK against normalised SNR with turbulence for a Rytov variance of $\sigma_I^2 = 0.5$. 94

Figure 6-14: Eye diagram representations of 4-PPM for a 200 mV$_{p-p}$ input signal with turbulence ($\sigma_I^2 = 0.1$) 95

Figure 6-15: Eye diagram representations of 4-PPM for a 400 mV$_{p-p}$ input signal with turbulence ($\sigma_I^2 = 0.1$) 96

Figure 6-16: Eye diagram representations of the received BPSK signal with turbulence ($\sigma_I^2 = 0.1$), a data rate of 50 Mbit/s, and a 400 mV$_{p-p}$ input signal. 96

Figure 7-1: (a) Block diagram of the experiment setup, and (b) the laboratory controlled atmospheric chamber and FSO link setup. 103

Figure 7-2: A plot of the Q-factor against the transmittance T. 105

Figure 7-3: The measured eye diagram of the received BPSK signal with (a) no fog conditions, and a data rate of 25 Mb/s, (b) fog attenuation at the transmitter side, and (c) fog attenuation at the receiver side (x-axis in ns), and (d) OOK received signal eye diagram for different P_{opt} levels (time scale is 20 ns/div), adopted from [180]. 107

Figure 7-4: Principles of the beam reflections between the convex and concave mirrors. 109

Figure 7-5: Pictures of the reflection spots showing the laser beam divergence at the receiver side Rx with: (a) no fog, (b) moderate fog, and (c) dense fog. 109

Figure 7-6: Block diagram of a terrestrial FSO link. 113

Figure 7-7: Q-Factor vs. link distance (km) at a 2 km visibility limitation factor for 0.65 μm, 1.55 μm, and 10 μm wavelengths. 116

Figure 7-8: The measured eye diagram of the received QPSK signal at the Rx side at: (a) 0.65 μm wavelength, (b) 1.55 μm wavelength, and (c) 10 μm wavelength (x-axis in ns). 117

Figure 7-9: (a) the laboratory controlled atmospheric chamber and FSO link setup, (b) the receiver end of the setup, and (c) a schematic of the fog chamber and FSO link setup. 122

Figure 7-10: A screenshot of the virtual instrument script created in NI LabVIEW. .. 122

Figure 7-11: The measured Q-factor values for the Hybrid 4-PPM-BPSK, 4-PPM and BPSK-SIM received signals at the same P_T and 5Mbit/s data rate for different T link values and fog conditions. 124

Figure 7-12: The measured eye diagrams of the received Hybrid 4-PPM-BPSK-SIM signal at a data rate of 25 Mb/s with: (a) no fog and $V = 320\ m$, and (b) fog and $V = 100\ \text{m}$. 125

Figure 7-13: The measured eye diagrams of the received 4-PPM signal at a data rate of 25 Mb/s with: (a) no fog and $V = 320\ \text{m}$, and (b) fog and $V = 100\ \text{m}$ 126

Figure 7-14: The measured eye diagrams of the received BPSK-SIM signal at a data rate of 25 Mb/s with: (a) no fog and $V = 320\ \text{m}$, and (b) fog and $V = 100\ \text{m}$ 127

LIST OF TABLES

Table 2-1: Optical sources. 22
Table 2-2: The gas elements of the atmosphere [83]. 24
Table 2-3: Photodetectors for FSO applications. 28
Table 2-4: Classification of lasers according to the IEC 60825-1 standard. 33
Table 3-1: The dependence of the spatial coherence length on wavelength and link distance for $C_n^2 = 10^{-12} m^{-2/3}$ 46
Table 3-2: The dependence of the spatial coherence length on wavelength and link distance for $C_n^2 = 10^{-15} m^{-2/3}$ 46
Table 5-1: Main parameters of the designed lab-chamber. 64
Table 5-2: Measured temperatures over six experiments at four different positions within the chamber. 69
Table 6-1: Turbulence strengths and the corresponding Rytov parameter. 80
Table 6-2: Main parameters of the turbulence chamber. 80
Table 6-3: Parameters of the FSO link demonstration. 82
Table 6-4: Measurement results of the turbulence strength inside the chamber. 84
Table 6-5: The simulation parameters. 92
Table 6-6: Experimentally measured Q-Factor values. 97
Table 7-1: Measured T and visibility values. 104
Table 7-2: Experimentally measured values for a 0.6 dBm transmitted signal. 108
Table 7-3: Parameters of the QPSK FSO link simulations. 113
Table 7-4: Atmospheric attenuation in different weather conditions for various wavelengths. 115
Table 7-5: Parameters of the FSO link with various modulations schemes. ... 119
Table 7-6: Measured T and visibility values. 123
Table 7-7: The experimentaly measured values for the 0.6dBm transmitted signal BER. 124

GLOSSARY OF ABBREVIATIONS

ADSL	Asynchronous digital subscriber loop
AEL	Accessible emission limits
AM	Amplitude modulation
ANSI	American National Standards Institute
APD	Avalanche photodiode
ASK	Amplitude Shift Keying
AWGN	Additive white Gaussian noise
BER	Bit-error-rate
Bps	Bits per second
BPSK	Binary phase shift keying
CDMA	Code division multiple access
CW	Continuous wave
DC	Direct current
DSL	Digital subscriber loop
EMC	Electro Magnetic Compatibility
EMI	Electromagnetic interference
ESA	European Space Agency
FCC	Federal Communications Commission
FT	Fourier transform
FIR	Far infrared
FHSS	Frequency-hopping spread-spectrum
FOV	Field of view
FSO	Free space optics

FSK	Frequency-shift keying
FTTH	Fibre to the home
GaAs	Gallium arsenide
GbE	Gigabit Ethernet
He Ne	Helium–neon laser
H-V	Hufnagel-Valley model of index of refraction structure parameter
IC	Integrated circuit
IEC	The International Electrotechnical Commission
IEEE	The Institute of Electrical and Electronics Engineers
IM/DD	Intensity-modulation/direct-detection
IID	Independently and identical distributed
IR	Infrared
ISS	Integrated Sensor Suite
IWS	The Intelligent Weather Station
ITU	The International Telecommunication Union
LAN	Local area network
LED	Light emitting diode
LMDS	Local multipoint distribution service
LOS	Line-of-sight
MIMO	Multiple-input multiple-output
MLCD	Mars laser communication demonstration
MMW	Millimetre-wave
MPE	Maximum permissible exposure
MZI	Mach-Zehnder interferometer
NASA	National Aeronautics and Space Administration
NEC	Nippon Electric Company
NEN	The Nearest-Neighbours

NIR	Near-infrared
NRZ	Non-return-to-zero
OBPF	Optical band pass filter
Ofcom	Office of Communication
OOK	On-off keying
OOK-NRZ	Non-return-to-zero OOK
OR_x	Optical receiver
OWC	Optical wireless communications
PAM	Pulse amplitude modulation
Pdf	Probability density function
PIN	PIN diode
P_{out}	Outage probability
PPM	Pulse position modulation
PSD	Power spectral density
QPSK	Quadrature phase shift keying
RF	Radio frequency
ROA	Real outdoor atmosphere
ROF	Real outdoor fog
RT	Receiver telescope
RMS	Root mean square
S.I.	Scintillation index
SILEX	Semiconductor-laser inter-satellite link experiment
SIM	Subcarrier intensity modulation
SNR	Signal-to-noise ratio
TIA	Transimpedance amplifier
TT	Transmitter telescope
UWB	Ultra wide band

WDM   Wavelength division multiplexing

GLOSSARY OF SYMBOLS

$A_0(r)$	Field amplitude as a function of position without atmospheric turbulence
A_d	Photodetector area
$\hat{A}$	Normal value
α	Effective number of atmospheric turbulence large scale eddies
α_e	Extinction ratio
B	Bandwidth in Hz
b	The pseudo-random sequence of data bits
B_e	Electrical filter bandwidth
β	Effective number of atmospheric turbulence small scale eddies
β_λ	attenuation coefficient
$\beta_a(\lambda)$	Wavelength dependent aerosol scattering coefficient
$\beta_m(\lambda)$	Wavelength dependent molecular scattering coefficient
C_n^2	Index of refraction structure parameter
c_k	Codeword
d	Photodetector diameter
$d(t)$	Pre-modulated data
D	Receiver telescope aperture
d_j	Signal level
D_T	Transmitter aperture
D_R	Receiver aperture
d_s	Optical source diameter
δ	Gaussian distributed phase fluctuation

$\vec{E}$	Electric field
$E[.]$	Expected value
E_g	Energy band gap in (eV)
f	Frequency in Hz
f'	Effective focal length
f_c	Carrier frequency in Hz
$g(t)$	Pulse shaping function
g	Photodetector gain/multiplication factor
(g^2)	Mean square current gain of an APD
G_R	Receiver effective antenna gain
G_T	Transmitter effective antenna gain
$\bar{g}$	Average APD gain
$\Gamma(\cdot)$	Gamma function
$\Gamma_x(\rho)$	Spatial coherence as a function of spatial distance
H	Humidity
$\hbar$	Altitude in metres
I	Received irradiance at the receiver plane
(i)	Mean generated electric current over a given period of time
$i_D(t)$	Subcarrier signal coherent demodulator output as a function of time
i_d	Photodetector dark current
i_{th}	Received signal threshold level
$i(t)$	Instantaneous current generated by a photodetector
$I(t)$	QPSK inphase channel
I_0	Mean received irradiance without atmospheric turbulence
I_s	Inner scale of turbulence

Symbol	Description
I_{sky}	Background irradiance from the sky
I_{Sun}	Background irradiance from the Sun
I_{peak}	Peak received irradiance
I_{bg}	Background radiation
$\langle i_d \rangle$	Mean photodetector dark current
k	Spatial wave number
K_{Bg}	Average photon count due to the background radiation
$K_n(\cdot)$	Modified Bessel function of 2nd kind and order n
K_s	Average photoelectron count per unit pulse position modulation time slot
k_B	Boltzmann constant
L	Link range
L_{Geom}	Geometric loss
L_M	Link margin
L_{op}	Optical/window loss
L_P	Pointing loss
L_T	Transmitter pointing- loss factor
L_R	Receiver pointing- loss factor
L_{slot}	Slot rate
λ	Wavelength
$\Delta\lambda$	Optical band pass filter bandwidth
Λ	Likelihood function
m	Power margin needed to achieve a given outage probability in dB
M	Number of signal levels or symbols
$m(t)$	Subcarrier signal
μ_t	Mean of sum of log-normal variables

N		Number of subcarriers
$\langle n \rangle$		Mean received photoelectron count
n		Atmospheric channel refractive index
N_0		Double side spectral density of the Gaussian noise
$N(\lambda)$		Spectral irradiance of the sky
n_b		Photoelectron count due to background radiation of power P_{Bg}
$n(t)$		Time-varying additive white noise
O_s		Outer scale of turbulence
P		Atmospheric pressure
$p(.)$		Probability density function
$P(t)$		Instantaneous received optical power
$\overline{p}(t)$		Unit energy pulse shape
P_0		Received average optical power in the absence of scintillation
P_{av}		Average transmitted optical power
P_{Bg}		Background radiation power
P_e		Unconditional bit error rate
P_e^*		Threshold or reference bit error rate
P_{ec}		Bit error rate conditioned on the received optical power/irradiance
P_{opt}		Optical power
P_R		Received optical power over a period of time
P_T		Transmitted peak optical power over a period of time
q		Electric charge
Q		Q factor
$Q(\cdot)$		Gaussian Q-function
$Q(t)$		QPSK quadrature channel

r		Position vector
R		Photodetector responsivity
R_b		Data rate
R_L		Equivalent load resistance
R_F		Fresnel zone
R_x		Receiver
ρ_0		Spatial coherence distance of a field traversing a turbulent channel
σ^2		Total noise variance
σ^2_{IMD3}		Third order inter-modulation distortion variance
σ^2_{Bg}		Background radiation noise variance
σ^2_{Dk}		Dark current noise variance
σ^2_I		Irradiance fluctuation variance
σ^2_N		Normalised irradiance fluctuation variance
σ^2_{Qtm}		Quantum noise variance
σ^2_{Sh}		Shot noise variance
σ^2_{Th}		Thermal noise variance
σ^2_l		Rytov variance
σ^2_x		Log-amplitude variance
T		Transmittance
T_b		Bit duration
T_{ps}		Pulse duration/symbol duration
t		Instantaneous time
T_e		Temperature in Kelvin
T_{fog}		Transmittance with fog
T_s		Pulse position modulation slot duration

T_{samp}	Sampling time
T_x	Transmitter
η_T	Transmitter radiation efficiencies
η_R	Receiver radiation efficiencies
θ_L	Optical local oscillator field phase
θ_c	Incoming optical carrier field phase
θ_s	Optical source divergence angle in milliradians
θ_T	Transmitter pointing-errors
θ_R	Receiver pointing-errors
V	Visibility
$v(r)$	Local wind velocity as a function of position, r
W	Word preamble
$W(\lambda)$	Spectral radiant emittance of the Sun
w_c	Intermediate RF frequency
$\langle x \rangle$	Mean value of variable x
x_0	Particle size parameter
$x_0(k)$	k-th transmitted symbol
$x_1(k)$	k-th transmitted symbol
$\gamma_T(\lambda)$	Total attenuation coefficient in m as a function of wavelength
$\tau(\lambda, L)$	Total extinction coefficient
$\gamma(a, x)$	Incomplete gamma function with variables a and x
$\gamma(I)$	Electrical SNR as a function of received irradiance
ξ	Optical modulation index
$\phi(r)$	Field phase as a function of position in an atmospheric turbulence channel

$\Phi_0(r)$	Field phase as a function of position without atmospheric turbulence
$\Phi_n(K)$	Refractive index fluctuation power spectral density as a function of K
$\varnothing$	Wavefront phase
X	Gaussian distributed log-amplitude fluctuation
$\Psi(r)$	Natural logarithm of the propagating field $E(r)$

DEDICATIONS

To my parents, my wife, my little daughter, and to the memory of my grandfather Ahmed Gatri

1. INTRODUCTION

1.1. Background

Free-space optical communication (FSO) has, since several centuries ago, been realised through the transmission of information-loaded optical radiation from a transmitter (T_x) to a receiver (R_x) over an atmospheric carrier. In military contexts, as long ago as around 800 BC, optical communication in the form of beacons was utilised to enable the transmission of messages to reach a receiver promptly [1, 2]. The weakness of this method was that only a limited number of predetermined messages could be conveyed, thus resulting in low information capacity [3]. Since then, FSO has continued to be researched and used principally by the military for the purpose of achieving secure communications. FSO has also been heavily researched for deep space applications by the National Aeronautics and Space Administration (NASA) and the European Space Agency (ESA), with programs such as the Mars Laser Communication Demonstration (MLCD) and the Semiconductor-Laser Inter-Satellite Link Experiment (SILEX) [4, 5, 6].

Experimental development of optical devices for high-speed communication over long distances demands a monochromatic and strong narrow optical beam at the desired wavelength. This would not be possible without the invention of the ruby laser in 1960 by Theodore Maiman, which is considered to be the first successful optical laser. In addition, the invention of the semiconductor optical lasers by Robert Hall in 1962 led to a significant improvement in FSO systems' reliability [7].

A television signal was broadcast over a 30-mile (48 km) distance using a gallium arsenide (GaAs) based light emitting diode (LED) by researchers working at the Massachusetts Institute of Technology (MIT) Lincolns Laboratory in 1962 [7]; while a record 118 mile (190 km) transmission of voice modulated Helium–neon laser (He-Ne) between Panamint Ridge and San Gabriel mountain, USA, was achieved in May 1963 [1]. The first laser link to handle commercial traffic was built in Japan by the Nippon Electric Company (NEC) in 1970. The link featured a full duplex 0.6328 µm He-Ne laser FSO between Yokohama and Tamagawa; a distance of 14 km [7, 8]. The research in FSO was proceeded to increase the system capacity as well as the link range and was majorly used by the military for covert communications in network-centric operational concepts that boosted the use of information as well as gaining superiority on the battlefield [10].

As a commercially complementary technology to the radio frequency (RF) (3 kHz to 30 GHz) and the millimetre-wave (MMW) (30 GHz to 300 GHz) wireless systems, FSO has now emerged as a reliable communication technology with rapid deployment of voice, data, and video within the access networks [11-13]. Data rates offered by wireless networks based on RF and MMW vary from a few megabits per second up to several hundred megabits per second depending on the system configurations (i.e. point-to-multipoint or point-to-point) [11, 14-15]. However, there is a limitation to theses systems' market penetration due to spectrum congestion, licensing issues, and interference from unlicensed bands. The availability of future license-free bands – referring to the frequency bands for which no exclusive licenses are required, and on which unregistered users can operate wireless devices without specific user or device authorizations, as per the 900 MHz, 2 GHz, and 5 GHz systems – is promising [16], but they still have some bandwidth and range limitations compared to FSO systems. The short-range FSO links are used as an alternative to RF links for the first or last mile to provide broadband access network to homes and offices as well as a high bandwidth bridge between the local and wide area networks [17].

Early doubts about FSO link performance, its dwindling acceptability by service providers, and slow market penetration, which were wide spread in the 1980s, are nowadays being challenged by service providers, organisations, and private establishments who are integrating FSO into their network infrastructure [9, 18].

Terrestrial FSO has now proven to be a feasible, complementary technology in addressing contemporary communication challenges, most particularly the provision of bandwidth/high data rates to end users at a low-cost. The fact that FSO technology is transparent to traffic types and data protocols makes its integration into existing access networks more rapid, reliable, and profitable compared to traditional fibre communications technology [19]. Despite these advantages, FSO performance is degraded by significant optical signal losses, due to the presence of aerosols, fog, smoke, and other suspended particles and gases in the atmosphere, which absorb and scatter the propagating optical and infrared waves, since their wavelengths are very close to the wavelengths of these frequencies [20]. Fog, compared to other atmospheric factors, is the dominant source for optical power attenuation, hence potentially reducing FSO link availability. However, in clear weather conditions, theoretical and experimental studies have shown that atmospheric turbulence (i.e. scintillation) can also gravely degrade the reliability and connectivity of FSO links [21]. As such, the atmospheric channel effects such as thick fog, smoke, and turbulence present a great challenge in achieving high link availability and reliability according to the Institute of Electrical and Electronics Engineers "IEEE" and the International Telecommunication Union (ITU), who specify a link availability standard of 99.999% (five nines) for the last mile access communication network. Hence, these channel effects still need to be studied and understood in order to enhance the link range and availability in terrestrial FSO systems [22-23].

1.2. Research Motivation and Justification

The increasing bandwidth requirements in communication systems is the driving force behind the current research in optical communications (both fibre optics and optical wireless). Optical communication guarantees abundant bandwidth, which equates to high data rate capabilities. The huge bandwidth available on the fibre ring networks that form the backbone of modern communication technology is still not available to end users within the access network. This is primarily due to the bandwidth limitations of the copper wire-based technologies that, in most places, liaise the end users to the backbone. This places a restrictive limit on the data rate/download rate available to end users. A number of solutions, including fibre to the home (FTTH), ultra-wide band technologies (UWB), digital subscriber loop (DSL) or cable modems, power-line communication, local multipoint distribution service (LMDS), and terrestrial free space optics (FSO), have been proposed to tackle this bottleneck.

The FTTH technology can deliver up to 10 Gbps to end users, and with the implementation of schemes such as dense wavelength division multiplexing (WDM), higher data rates of up to 10 Tbps are achievable [15, 26]. The most important challenge facing large deployment of the FTTH technology is its prohibitive cost of implementation. Furthermore, both DSL and power-line communication systems are copper-based technology which makes them less attractive [27]. The UWB technology uses the unlicensed radio spectrum in the 3.1 – 10.6 GHz band for short-range communication [16, 28]. It is also a copper wire based solution, and its potential data rates are at odds with several Gbps available in the backbone. The interference of UWB signals with other systems operating within the same frequency spectrum is another drawback [29, 30, 115].

The FSO technology presents a number of advantages over other technologies, especially RF. One of the most important advantages is the immunity of FSO to electromagnetic interference (EMI). The very narrow optical signal offers a secure link with enough spatial isolation from potential interferers, which makes it the favoured option in certain applications where there is a requirement for a very low level of interference or no interference at all [19]. Moreover, FSO has another advantage over RF— namely the increased security afforded by the laser's small footprint, which makes detection, interception and jamming very difficult. Further advantages provided by FSO over RF include the possibility of rapid deployment, and the flexibility of setting up temporary communication links with higher data rates, low cost, small size, and limited power usage. Furthermore, FSO links consume low power, offer better security against electromagnetic interference, do not require licensed spectrum, and offer low system and installation costs [24, 25].

FSO is a fibreless, laser or light-emitting diode (LED) driven technology, which provides similar capacity to that of optical fibre based communications with a substantial reduction in cost and time [62]. The integration of FSO into the access network can be done inexpensively and rapidly, as it is transparent to traffic type and protocols. However, the channel in FSO is the atmosphere, which presents a great challenge to the performance of an FSO system since it is subject to various impairments that arise in the atmosphere. Hence, it is essential to experimentally characterise the different atmospheric conditions such as smoke, fog and turbulence, as well as analyse the system performance under these atmospheric circumstances [31, 32].

A number of researchers have examined the effect of atmospheric turbulence [33, 34]. However, most of these studies are theoretical, and relatively little work has been reported experimentally. This is because, in practice, it is very challenging to measure the effect of atmosphere fading under diverse real conditions. This is primarily due to the long waiting time required to observe and experience a reoccurrence of different atmospheric cases, as well as the requirement for complex and costly measuring tools and equipment. Aside from attenuation, another crucial channel effect is atmospheric scintillation. It is caused by atmospheric turbulence, which is a direct product of the atmospheric temperature inhomogeneity [35, 36]. The scintillation effect results in signal fading due to the constructive and destructive interference of the optical beam crossing the atmosphere. For FSO links spanning 500 meters or less, typical scintillation fade margins are 2 to 5 dB, which is less than the margins for atmospheric attenuation [37], making scintillation unimportant for short-range FSO systems.

Several methods can be used to overcome the effect of turbulence, such as using a multiple input multiple output (MIMO) system [40, 41], utilising temporal and spatial diversity [34], or aperture averaging [42]. Nevertheless, selecting a modulation format that is most immune to scintillation effects is also important [43, 44].

By evaluating the performance of FSO systems in turbulent atmospheric channels, researchers have analysed the bit error probabilities based on on-off keying (OOK) modulated FSO systems and overall turbulence regimes [47, 51, 52]. The OOK modulation technique is very simple to apply, and is, consequently, the current modulation technique of choice in all commercially available terrestrial FSO systems [27,

51, 53, 54]. In a standard OOK-FSO system, the threshold level/point in the decision circuitry used to differentiate between bits '0' and '1' is constantly fixed midway between the expected levels of data bits '0' and '1' [55]. Under turbulence induced channel fading, the received signal level fluctuates. This means that the threshold detector has to track this fluctuation to find out the optimum decision point [57]. This presents a great design challenge as the channel noise and fading will have to be regularly tracked for the OOK-FSO system to perform optimally.

In order to address the symbol detection challenges of OOK-FSO in turbulence affected channels, this research investigates the binary phase shift keying pre-modulated subcarrier intensity modulation (BPSK-SIM) and the pulse position modulation (PPM) as possible alternative modulation techniques. Furthermore, in this book, the error performance of SIM-FSO is broadly investigated in the following atmospheric turbulence regimes – weak, moderate and strong attenuation channels.

In this book, an experimental study investigating the effects of temperature induced turbulence on FSO link performance for the BPSK-SIM and PPM data formats is carried out. The research also study's the dependence of the data transmission performance against the turbulence source position along the link.

Apart from turbulence, fog has a prominent impact on FSO links; limiting the link range to a few hundred metres under heavy fog conditions [38]. However, when the link length exceeds several hundred metres, irradiance fluctuations of the received optical signal due to turbulence poses a critical problem. The turbulence induced by the random fluctuations of temperature and pressure leads to a random variation of the atmospheric refractive index. The variations in the refractive index along the optical paths in turn cause random fluctuations to the received optical irradiance, which results in severe system performance degradation [39]. However, few experimental studies on the BER and Quality Q Factor performance of an FSO link in fog affected channels are reported. A number of research studies investigating the effect of fog on FSO systems largely concentrate on the measurement of attenuation and channel modelling [38, 45, 46]. Therefore, this research aims to investigate and mitigate the effect of fog by applying power efficient modulation schemes.

Furthermore, an experimental study of the fog effect on FSO links performance for the binary-phase-shift-keying coherent detection (BPSK) data format is conducted, and the experimental results for fog induced attenuation for measured visibility of $V < 0.5$ km are reported. Optical solutions to mitigate fog fading are reported, and the improvement of the beam spot size through applying a combination of concave and convex mirrors, which consequently improves the link performance, is studied. Different methods are introduced and demonstrated in order to control the different turbulence levels and to determine the performance equivalence between indoor and outdoor FSO links. The effect of turbulence and fog on FSO communication systems for 4-PPM, BPSK-based SIM, as well as hybrid BPSK-SIM-PPM modulation schemes, is experimentally investigated and evaluated. The following research objectives were devised in order to achieve these goals.

1.3. Research Objectives

The prime aim of this work is to experimentally characterise and investigate the atmospheric channel effects, especially fog and turbulence, on the performance of FSO links. Hence, a dedicated laboratory chamber of 3-metre length with inlets and outlets to simulate and demonstrate the atmospheric effects on an FSO channel in a controlled environment was designed and built.

In this specific laboratory chamber, an advanced wireless weather station was utilised to provide real-time atmospheric values for temperature, humidity, barometric pressure, wind speed and direction, dew point, and rainfall. In order to have an ideal experimental environment, the chamber was built on an optical table in the Electro Magnetic Compatibility (EMC) Chamber at the Bonn-Rhein-Sieg University of Applied Science Communication laboratory.

A list of research objectives has been defined in order to accomplish these aims:

- Review the fundamental characteristics of terrestrial FSO links and understand the properties of the atmospheric channel and the challenges affecting the system performance.

- Review models for describing the channel fading induced by the turbulence and fog-induced atmospheric effects, in order to understand the limits and range of validity of each model.

- Investigate and compare different empirical fog models available in literature for different parameters such as visibility V (km), transmittance threshold T-th, and wavelength λ.

- Investigate the performance of BPSK-SIM, 2-PPM, 4-PPM and the hybrid BPSK-SIM-PPM modulated FSO systems over a turbulent channel.

- Experimentally verify and characterise the atmospheric turbulence and calibrate it to a real outdoor FSO channel.

- Experimentally investigate fog effects on the transmitted optical beam over an FSO channel, and compare the system performance of BPSK-SIM-FSO and 4-PPM-FSO under fog conditions.

- Investigate, by means of simulation, the performance of QPSK-FSO at various wavelengths under clear and fog-affected conditions.

- Investigate hybrid BPSK-SIM-PPM modulated FSO systems over a channel with fog-induced attenuation.

1.4. Book Organisation

This book is divided into eight chapters. Following the introduction chapter, **Chapter Two** previews the basic fundamentals of FSO systems. In this chapter, a complete overview of FSO technology is introduced, including its specific features and applications. The general block diagram of an FSO communication system is presented, and the functions of each part are highlighted. Eye safety issues are discussed in the remainder of the chapter.

In **Chapter Three**, a review of atmospheric turbulence is presented, and its effect on optical radiation over the atmosphere is discussed. The three different statistical distributions for modelling the irradiance fluctuations of optical radiation over turbulent atmosphere are investigated. The discussion of these models is crucial as they are used in later chapters to describe the statistical behaviour of the received signal in all ranges of FSO links.

In **Chapter Four,** we discuss and investigate the commonly used FSO modulation techniques and their limitations and performance under atmospheric fading channels.

Alternative robust modulation schemes are also studied to investigate the potential increase in efficiency that can be achieved in terms of power and bandwidth.

In **Chapter Five,** a detailed description of the design of the atmospheric chamber is introduced. Furthermore, the methods to produce homogeneous fog and its calibration to the real outdoor fog (ROF), and the data acquisition for fog attenuation measurements are described in detail. In addition, various methods of producing and controlling atmospheric turbulence are presented. The experimental results of spectrum attenuation of FSO communication systems operating at visible and near-infrared (NIR) wavelengths (0.6 μm $< \lambda <$ 1.6 μm) are also presented. The results are compared with the selected empirical fog models in order to formalise their performance practically from dense to light fog conditions.

In **Chapter Six,** the performance of FSO links under controlled turbulence atmospheric channels is studied, and both BPSK-SIM and 4-PPM, as well as hybrid PPM-BPSK-SIM modulations techniques, are investigated.

In **Chapter Seven,** different aspects of loss mechanisms encountered in the design of an FSO system, including channel absorption and scattering, are examined. The link budget expression, as well as the estimation of achievable link ranges and margins, are also covered in this chapter. The experimental set-up of the fog-affected chamber used to study the performance of various modulation techniques is presented. The experiment to assess BPSK-SIM-FSO, 4-PPM and hybrid PPM-BPSK-SIM system performance is performed using an indoor open atmospheric fog-affected chamber (3-6 m in length). The transmission distance is increased to 14.7 m by means of multiple reflections of the laser beams between mirrors. Moreover, the effect of fog over an FSO communications system employing a QPSK modulation format and operating at various wavelength regimes is investigated.

In **Chapter Eight,** a summary of the key findings is presented, and the conclusions and future work are discussed.

1.5. Original Contributions

As a direct result of this research, the following original contributions have been made:

- The design of a dedicated laboratory atmospheric chamber within an intelligent weather station with a feedback loop. The chamber is used to assess the effect of temperature induced turbulence on FSO links performance in real time and to enable several measurements over a long period of time (see Chapter Five). Furthermore, methods to generate a different levels of turbulence, and its control, are highlighted and practically implemented in Chapter Six. Atmospheric turbulence is characterised and calibrated using the atmospheric chamber, and the experimental measurements confirm the statistical based log-normal model. The relationship between the indoor environment and the outdoor FSO link is also obtained in order to ensure total reciprocity of atmospheric turbulence.

- The performance of BPSK-SIM and 4-PPM FSO links under a controlled turbulence environment are numerically and experimentally investigated in Chapter Six. The numerical simulation results for the bit error rate (BER) performance of the PPM-BPSK modulation scheme in the log-normal atmospheric channel for a range of turbulence variance are presented. Results in Chapter Six show that PPM-BPSK offer a similar performance to 2-PPM, and a superior performance compared to BPSK-SIM while having the same bandwidth. However, an inferior performance compared to 4-PPM is observed for the same turbulence levels.

- An experimental evaluation of the performance of coherent detection BPSK modulation schemes under the effect of fog for FSO communication links is carried out in a controlled laboratory test-bed. The effects of low to high visibility on the FSO link BER performance in the presence of fog is also measured and investigated in Chapter Seven. The dependency of the FSO link performance on the fog intensity variation, as well as the source placement of the fog induction over the channel, is studied. Moreover, an experimental study and comparison between BPSK-SIM and 4-PPM, as well as the newly proposed Hybrid 4-PPM-BPSK scheme under a controlled fog chamber and an indoor open chamber are presented in Chapter Seven. Error probabilities and Q values, as well as eye diagrams of aforementioned modulation techniques, are presented.

- Simulation results of a QPSK FSO system at various operating wavelengths are introduced in Chapter Seven. The power losses on the R_x side are analysed, together with the link performance in terms of the received power level and the Q factor under clear and fog-induced atmospheric attenuation.

The overall contribution of this book is schematically illustrated using a research roadmap in Figure 1-1.

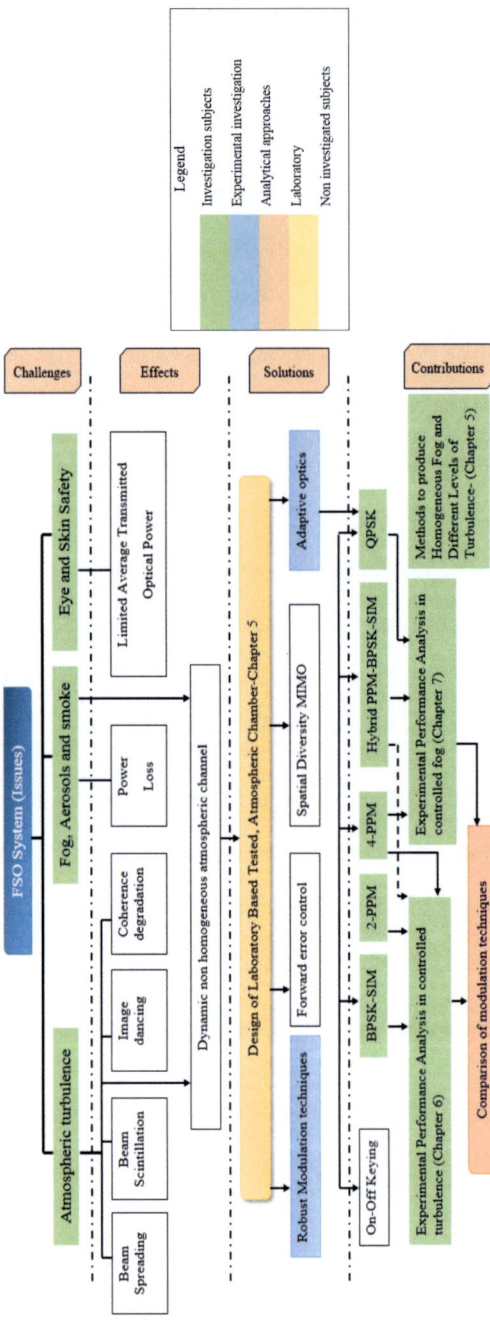

Figure 1-1: A schematic illustrating the research roadmap and the original contributions.

2. FUNDAMENTALS OF FSO

2.1. Introduction

FSO communications is a technology that utilises visible and infrared (IR) light propagating through the atmospheric channel to transmit information. FSO is a complementary technology to both RF and optical fibre communication systems. It has been proposed as a 'last mile' solution in broadband wireless access links in certain applications and scenarios where RF based technologies may not be able to be adopted because of the shortage of available frequency spectrum [58]. In this chapter, the fundamentals of FSO technology, such as its applications, features, and terminology are presented. The chapter outlines the optoelectronic devices utilised in FSO systems and concludes with a discussion on the eye/skin safety requirements associated with laser transmission.

2.2. Overview of FSO Technology

FSO technology consist of the transfer of data/information between two points using optical radiation as the carrier signal through unguided channels. The transported data is modulated on the intensity, phase, or frequency of the optical carrier. An FSO link is fundamentally based on line-of-sight (LOS) transmission with no obstruction along the propagation path, thus ensuring very high data rate transmission in the absence of any multipath induced dispersion [59, 60]. The unguided channels could be any, or a combination of, space, sea-water, or atmosphere. The focus in this work is on terrestrial FSO and, as such, the channel of interest is the atmosphere.

FSO systems allow the use of narrow divergence, directional laser beams, which provide high security with a low probability of interception or detection; features that can be desirable in many applications. Narrow and highly focused optical beams are also important in situations where the channel is not clear due to fog. For example, penetration of dense fog over a kilometre distance is achievable at Gbps data rates with a beam divergence of 0.1 mrad. Furthermore, the tight antenna patterns of FSO links allow considerable spatial re-use, and wireless networks using such connectivity are highly scalable, in marked contrast to ad-hoc RF networks, which are non-scalable [61, 62].

The basic features, areas of applications, and the description of each fundamental block of an FSO system are further discussed in the following sections.

2.3. Features of FSO

Several typical features of FSO technology are presented below:

i. ***Huge modulation bandwidth*** – The frequency range of an optical carrier spans from 10^{12}-10^{16} Hz, which corresponds to 2000 THz bandwidth. This is a key feature since the amount of data that can be transmitted is directly related to the bandwidth of the carrier. Optical communication permits a far greater information capacity compared to RF technology, with an operational frequency bandwidth, which is relatively lower by a factor of 10^5 [60, 63, 64].

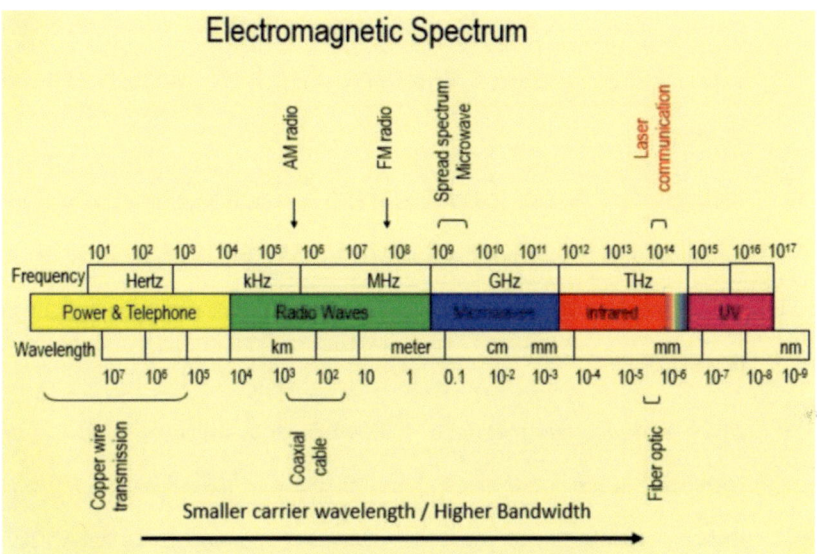

Figure 2-1: The overview of the EM spectrum with its nominated frequency bands [66].

ii. **Unlicensed spectrum** – FSO utilises the visible to far infrared (FIR) bands of the electromagnetic spectrum, as shown in Figure. 2-1. The optical carrier frequencies have a very small footprint at this spectrum range, and therefore, FSO systems are not significantly affected by signals from other bands. However, this is not the case with the RF spectrum, where adjacent bands are susceptible to interference. Therefore, regulatory authorities, such as the Federal Communication Commission (FCC) in the US, and the Office of Communication (Ofcom) in the UK, have put stringent regulations in place [67]. Obtaining a slice of the RF spectrum is costly and could take a long time. However, this is not the case with the FSO spectrum, which is free, relatively inexpensive compared to the RF spectrum, and can be rapidly installed.

iii. **Narrow beam size** – Optical radiation is well known for its extremely narrow beam size, which typically ranges between 0.01 and 0.1 mrad [65]. This implies that the transmitted power is concentrated within a very small area. The tight spatial confinement also permits parallel transmission of a number of laser beams at the same, and/or different wavelengths, which is not possible in RF-based systems.

iv. **Cost-effective** – The cost of utilisation of FSO systems is lower than that of an RF system for a given data rate. FSO can deliver similar bandwidth to optical fibres, but without the additional cost of right-of-way and trench digging, especially in urban areas [68].

v. **Quick to deploy and redeploy** – FSO offers quick deployment and can be operational in just a few hours, from installation to link alignment, which is desirable in emergency situations where there is an urgent need to establish a high bandwidth communications link. The most important requirement in FSO deployment is the achievement of an unobstructed LOS between the T_x and the R_x. FSO modules can also be taken down and redeployed in different locations [64, 69].

vi. **Weather dependent** – Atmospheric conditions have a huge impact on the performance of terrestrial FSO communication links due to the presence of atmospheric absorption and scattering. These 'undetermined' characteristics of the FSO channel undoubtedly pose the greatest challenge. Nevertheless, this is not particular to FSO, since RF and satellite communication links are also subject to link unavailability due to heavy rain, and snow [96, 123].

2.4. FSO Applications

Due to the features discussed above, FSO is considered to be an attractive technology for various applications. Below are a number of applications where FSO technology could be utilised [70, 71]:

Last mile access – FSO technology can be used to bridge the bandwidth gap, known as the last mile bottleneck, which exists between end users and the fibre optics backbone. Today's FSO-available products on the market feature links spanning from 50 m to a few km, and data rates ranging from 1 Mbps to 10 Gbps [27, 63, 64].

Cellular communication backhaul – FSO technology can be used to carry traffic between base stations and switching centres in third/fourth generation (3G/4G) networks. It can also transport the IS-95 code division multiple access (CDMA) signals from macro- and micro-cell sites to the base stations [27].

Backup to optical fibre links – FSO links can be used as backup links in conditions where optical fibre communication links are not available, broken or damaged [27, 72]. FSO is also generally utilised to conserve high link availability in cases where fibre optics links are practically difficult to use due to complex physical implementations [73].

Disaster recovery/temporary links – FSO links could be used as temporary links when existing communication networks have been damaged [74, 75, 76].

Multi-campus communication network – FSO has found applications in interconnecting campus networks and providing backup links at fast-Ethernet or gigabit-Ethernet speeds [77].

Difficult terrains – FSO technology can be used in situations where installing cable-based communication links is challenging or costly.

High-definition television – In broadcasting live events from remote areas and war zones, signals from the cameras have to be sent to the broadcasting vehicle, which is connected to a central office via a satellite uplink. In such scenarios, high-quality transmission between the camera and the vehicle can be offered via FSO technology [1, 60, 63, 64].

2.5. The FSO System

The block diagram of a typical terrestrial FSO system is presented in Figure 2-2. In common with other communications technologies, an FSO system fundamentally consists of a T_x, a channel, and an R_x.

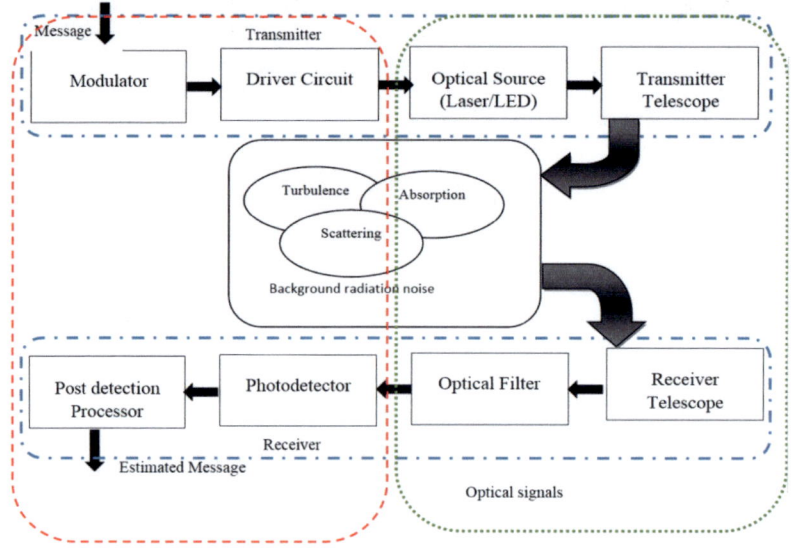

Figure 2-2: Block diagram of a terrestrial FSO system.

2.5.1. The Transmitter

The Tx's main role is to convert an electrical signal into an optical signal using a laser or an LED. The optical signal is then transmitted to the R_x over the atmosphere. The T_x is typically comprised of a modulator, a driver, a laser diode and transmitting optics. The most commonly used modulation type is the IM, in which the data source is modulated onto the intensity of the optical radiation. This is achieved by changing the drive current of the optical source in accordance with the data to be carried. Alternatively, an external modulator, such as a Mach-Zehnder interferometer (MZI), could be used to modulate the frequency or phase of the optical carrier signal [1]. Using an external modulator ensures a higher data rate than direct modulation, however, such modulators typically have a nonlinear response. A transmitter telescope (TT) is used for collimating the optical beam prior to transmission. Table 2-1 shows a summary of optical sources that are usually utilised in FSO systems.

Table 2-1: Optical sources.

Wavelength (nm)	Type	Features
~850	Vertical cavity surface emitting laser	• Cheap and readily available (CD lasers) • No active cooling • Low power density • Data rates up to ~10 Gbps • Typical power: 6 mW
~1300/~1550	Fabry-Perot lasers Distributed-feedback Lasers	• Low eye safety criteria • High power density (100 mW/cm^2) • Typical power: 28 mW. • Compatible with Erbium-doped fibre amplifiers (EDFA). • High speed, up to 40 Gbps. • Efficiency of 0.03–0.2 W/A. • Typical power can reach 1–2 W when merged with Erbium-doped fibre amplifiers [53].
~10,000	Quantum cascade laser	• Expensive. • Very fast and highly sensitive laser. • Components not readily available. No output power higher than 100 mW.
Near Infrared	LED	• Cheap • Non-coherent • Low power density, hence safer [78] • Simpler driver circuit • Low data rates: < 200 Mbps [79] • Typically low power: < 10 mW

Most FSO systems are designed to operate in the range between the 780–850 nm and 1520–1600 nm spectral windows. The first range is the most widely adopted due to the fact that suitable devices and components are readily available and at a low cost [27]. The 1550 nm band is, however, attractive for several reasons including a) compatibility with the third transmission window in optical fibre communications; b) achievability of a higher transmit optical power than the 780 nm band, and c) a lower solar background radiation level and scattering in light fog. Therefore, at 1550 nm, more power can be transmitted to surmount attenuation due to fog [80].

2.5.2. The Atmospheric Channel

The transmitted laser beam travelling through the atmosphere encounters absorption, scattering, and fluctuation due to atmospheric conditions such as snow, rain, fog, low cloud, turbulence, smoke, and dust particles [41, 81]. However, fog and turbulence are the two dominant atmospheric conditions that affect the FSO link reliability and availability [82, 83]. Optical communication channels are different from the conventional Gaussian noise channel, due to the fact that the signal $x(t)$ constitutes power rather than amplitude. Therefore, the two main constraints on the transmitted signal are: (a) $x(t)$ must be positive in value and (b) the average value of $x(t)$ must be less than the specified maximum power P_{max}, which is given by:

$$P_{max} \geq \lim_{T \to \infty} \frac{1}{2T} \int_{-T}^{T} x(t)d(t) \qquad (2.1)$$

The constituents of the atmospheric channel, which comprise of various gases and aerosols, are listed in Table 2-2. Aerosols have diverse nature, shapes, and sizes. Furthermore, aerosols can vary in terms of distribution, constituents, and concentration. In the presence of aerosols in the channel, the interaction between the aerosols and light can have a large dynamic in terms of λ.

Table 2-2: The gas elements of the atmosphere [83].

Constituent	Volume Ratio (%)	Parts Per Million ppm)
Nitrogen (N_2)	78.09	
Oxygen (O_2)	20.95	
Argon (Ar)	0.93	
Carbon dioxide (CO_2)	0.03	
Water vapour (H_2O)		40 – 40,000
Neon (Ne)		20
Helium (He)		5.2
Methane (CH_4)		1.5
Krypton (Kr)		1.1
Hydrogen (H_2)		1
Nitrous oxide (N_2O)		0.6
Carbon monoxide (CO)		0.2
Ozone (O_3)		0.05
Xenon (Xe)		0.09

2.5.3. Atmospheric Attenuation

Atmospheric attenuation is the process whereby some, or most of the electromagnetic wave energy is lost by traversing the atmosphere. The atmosphere causes signal degradation and attenuation in an FSO system in various ways, including absorption, scattering, and scintillation. All these effects vary with time and the current local conditions and weather. Generally, atmospheric attenuation is defined by the Beer-Lambert law as follows [27]:

$$\tau(\lambda, L) = \frac{P_R}{P_T} = \exp(-\gamma_T(\lambda)L), \qquad (2.2)$$

where $\gamma_T(\lambda)$ and $\tau(\lambda, L)$ represent the total attenuation/extinction coefficient (m^{-1}) and the transmittance of the atmosphere at a given λ respectively. P_T is the transmit optical power and P_R is the received optical power after a defined transmission span L.

Scattering is the process by which small particles suspended in a medium of a different index of refraction diffuse a portion of the incident radiation in all directions. With scattering, there is no energy transformation, but a change in the spatial distribution of the energy [48]. The scattered light is polarised, and of the same λ as the incident λ, which means that there is no loss of energy to the particle [84].

The three main types of scattering are: (1) Rayleigh scattering, (2) Mie scattering, and (3) non-selective scattering, as shown in Figure 2-3.

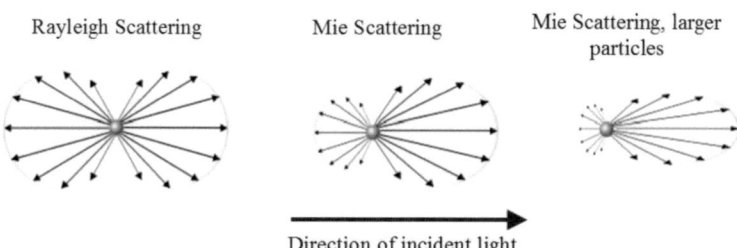

Figure 2-3: Comparison between Rayleigh, Mie scattering and Mie Scattering with larger particles [49].

Another challenging aspect of the FSO channel is atmospheric turbulence. Part of the radiation directed from the Sun towards the Earth is absorbed by the Earth's surface, thus causing the surface to heat up. The resulting mass of warm and light air mixed with the cooler air results in turbulence. This culminates in a small (in the range of 0.01 to 0.1 degrees) but spatially and temporally fluctuating atmospheric temperature [65]

The temperature inhomogeneities of the atmosphere lead to fluctuations in the refractive index that cause variations in the size of air packets from ~0.1 cm to ~10 m. These air packets behave like refractive prisms of diverse indices of refraction. The propagating optical radiation is, therefore, fully or partially deviated depending on the relative size of the beam and the degree of temperature inhomogeneity along its path. As such, the optical radiation passing through the turbulent atmosphere experiences random variation/fading in its irradiance (scintillation) and phase [53, 65, 85].

Atmospheric turbulence depends on wind speed, the variation of refractive index, and the atmospheric altitude/pressure. The impact of atmospheric turbulence on laser-based communications includes [86]:

a) *Beam steering* – angular deviation of the beam from its initial LOS target, which leads to the beam being out of the R_x aperture range.

b) *Image dancing* – due to variations in the beam's angle of arrival, the received beam's focus moves arbitrarily in the image plane.

c) *Beam spreading* – the beam divergence is increased because of the presence of scattering, which results in a reduction in the received power density.

d) *Beam scintillation* – the optical beam phase front is impaired by variations in the scintillation index, leading to irradiation fluctuations, or scintillations.

e) *Spatial coherence degradation* – the phase coherence throughout the beam phase fronts suffers losses due to turbulence [87].

f) *Polarisation fluctuations* – these result from changes in the state of polarisation of the received optical field after traversing a turbulent medium. However, Polarisation fluctuations are negligible for a horizontally travelling optical beam, as per several studies conducted to estimate the magnitude of the change of polarization in an electromagnetic signal as it travels through turbulent atmosphere [88, 207, 208]. Several expressions are used (with similar outcomes) to determine the depolarization of the electromagnetic field of a quasi-monochromatic light source at a particular optical frequency. For instance, for

L=1500m,

λ= 1550 nm, and C_n^2= 1x10^{-13}, the depolarized component is 2.1 × 10^{-18} smaller than the polarized component [209].

g) The modelling and statistical behaviour of the received irradiation, atmospheric fog and turbulent channel will be examined in detail in Chapter Three.

2.5.4. The Receiver

The transmitted data from the incident optical field is recovered and collected by the R_x, which consists of the following:

a) R_x *telescope* – is used principally to collect, focus and guide the incoming radiation towards the photodetector. Therefore, a large R_x telescope aperture is essential for collecting multiple non-coherent radiations.

b) *Optical bandpass filter* – is used to reduce the amount of background radiation.

c) *Photodetector* – positive intrinsic negative (p-i-n) diodes (PIN) or avalanche photodiodes (APD) are both suitable photodiodes, which convert the incident optical field into an electrical signal. The most commonly used photodiodes and their characteristics are summarised in Table 2-3.

d) *Post-detection processor/decision circuit* – the amplification, filtering and further signal processing are carried out in this subsystem in order to ensure high fidelity of the recovered data.

In this research, a Thorlabs PDA36A detector, which includes a reverse-biased PIN photodiode, mated to a switchable gain transimpedance amplifier, and packaged in a rugged housing (as sketched in [223]), was used.

Table 2-3: Photodetectors for FSO applications.

Material/Structure	Wavelength (nm)	Responsivity (mA/W)	Typical Sensitivity	Gain
Silicon PIN	300–1100	0.5	-34 dBm@155 Mbps	1
Silicon PIN, with transimpedance amplifier	300–1100	0.5	-26 dBm@1.25 Gbps	1
InGaAs PIN	1000–1700	0.9	-46 dBm@155 Mbps	1
Silicon APD	400–1000	77	-52 dBm@155 Mbps	150
InGaAs APD	1000–1700	9	-33 dBm@1.25 Gbps	10

The R_x detection process can be divided into two categories:

Direct detection receiver – is the most widely used receiver and has a rather simple implementation [86].

Coherent detection receiver – is based on the photo-mixing process. Coherent detection can be either homodyne or heterodyne. In a homodyne R_x, the frequency of the local (optical) oscillator is precisely the same as that of the incoming radiation [86]. In contrast, in heterodyne detection, the incoming radiation and the local oscillator frequencies are different [89]. The main advantages of a coherent R_x are easy amplification at an intermediate frequency and the improved signal-to-noise ratio, which is achievable through increasing the power of the local oscillator.

The R_x deployment and the detection process are discussed in more detail in the next chapter.

2.6. Noise in Optical Detection

There are several noise sources that are associated with optical communication systems, principally originating from the background radiation and the internal mechanisms behind optical and electronic devices [86, 89]. The dominant sources of noise are summarised below.

2.6.1. Thermal Noise

Thermal noise results from the interaction between the free electrons and the vibrating ions in a conducting medium [70, 86, 89]. It is known and viewed as 'white' noise due to the fact that the power spectral density (PSD) is frequency independent. The Gaussian distribution is used to describe the statistical behaviour of the thermal noise with zero mean and variance [79]:

$$\sigma_{Th}^2 = \frac{4k_B T_e B}{R_L}, \qquad (2.3)$$

where k_B is the Boltzmann's constant (1.3806503×10^{23} m^2 kg s^{-2} K^{-1}), T_e is the temperature in Kelvin, B is the bandwidth in Hz and R_L is the equivalent load resistance in Ohm (Ω) and is normalised to 1 Ω.

2.6.2. Photon Fluctuation Noise

An important source of noise affecting all types of photodetectors is related to the quantum nature of light itself, and is due to the varying number of photons provided by a coherent optical source (the mean radiation intensity remains constant) [86]. The variance of quantum noise can be expressed by [1, 70]:

$$\sigma_{Qtm}^2 = 2q(i)B, \qquad (2.4)$$

where q is the electric charge, and (i) is the mean generated electric current over a given period of time.

2.6.3. Dark Current Shot Noise

The dark current is the photocurrent produced when no photons are incident on the photodetector. The dark current, originating from the transition of electrons from the valence to the conduction band, includes a tunnel, leakage, and diffusion currents, as well as a generation-recombination process taking place in the space-charge region, which is proportional to the volume of the depletion region [135]. The value of the dark current is dependent on the energy band-gap of the photodetector material. The shot noise variance is given by [1]:

$$\sigma_{Dk}^2 = 2q(i)B. \tag{2.5}$$

2.6.4. Background Radiation

Background radiation noise results from the detection of photons caused by the environment. This type of noise is dominated by two main sources, namely localised point sources (e.g. the Sun) and extended sources (e.g. the sky). The background radiation from other sources including celestial bodies (such as stars) and reflected background radiation, are assumed to be too weak to be considered for a terrestrial FSO link. Nevertheless, such sources contribute significantly to background noise in deep space FSO. The impact of background noise can be usually decreased by using an optical band-pass filter (OBPF) and an R_x with a very narrow field of view (FOV). The background radiation considered as the shot noise has a variance of [69]:

$$\sigma_{Bg}^2 = 2qB\Re(I_{sky} + I_{Sun}),\tag{2.6}$$

where I_{sky} and I_{Sun} represent the radiation from the sky and the Sun respectively. Generally, the background radiation noise is greater than other noise processes [1, 86], resulting in dominating the whole shot noise, which is the sum of all types of noise.

2.7. Eye Safety and Standards

According to [27, 92], some high-power laser beams used for medical procedures can damage the human skin. However, the part of the human body most susceptible to serious laser damage is the eye. As per sunlight, laser light travels in parallel rays. Like staring directly into the Sun, exposure to a laser beam of sufficient power can result in a permanent eye injury [66, 93]. The eye can focus light covering $0.4 < \lambda < 1.4$ μm on the retina, while other wavelengths tend to be absorbed by the front part of the eye (the cornea) before the energy is focused. Figure 2-4 shows the absorption of light by the human eye at different wavelengths.

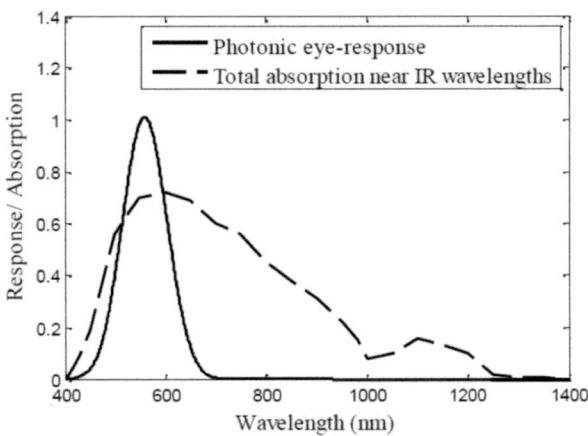

Figure 2-4: Response/absorption of the human eye at various wavelengths [53].

There are a number of international standard bodies which provide standards on safety for working with IR radiations. A list including some of such organisations and their standards is presented in [31, 90].

Table 2-4 presents the main characteristics and requirements for the various categories of the lasers classification system, as specified by The International Electrotechnical Commission (IEC 60825-1) standard [80, 91]. For instance, classes 2 and higher must have the triangular warning label, and other labels are required in specific cases signalling laser emission, laser apertures, skin hazards, and invisible wavelengths.

Table 2-4: Classification of lasers according to the IEC 60825-1 standard.

Category	Description
Class 1	Low power device with radiation at a wavelength range of 302.5–4000 nm. The device is intrinsically safe by its technical design under all reasonably predictable usage circumstances, including applying optical vision instruments (binoculars, microscopes, monocular).
Class 1M	Similar to Class 1, but with large diameter beams which are divergent. As such, there is a possibility of danger when viewed with optical instruments such as binoculars, telescope, etc.
Class 2	This class is safe and it is limited to a 1mW continuous wave (CW). Eye protection is normally ensured by the defence reflexes of the eye, including the palpebral reflex (closing of the eyelid). The palpebral reflex offers efficient protection over all foreseeable utilisation conditions, admitting vision using optical instruments (binoculars, microscopes, monocular).
Class 2M	Visible radiation from a low power device (in the 400–700 nm band). Eye protection is usually guaranteed by the defence reflexes provided by the palpebral reflex (closing of the eyelid). Similar to Class 2, the palpebral reflex offers an effective protection under all usage conditions, however, not in the case of using optical instruments (binoculars, microscopes, monocular).
Class 3R	Average power device emitting radiation in the range of 302.5–4000 nm. Direct vision is potentially dangerous.
Class 3B	Average power device emitting radiation in the 302.5–4000 nm band and has a power limit of 30mW. Direct vision of the beams is hazardous. Medical checks and specific training are required before installation or maintenance is carried out.
Class 4	As a higher power class, the device is always dangerous to the eye and the skin, and fire risks also exist. Must be equipped with a key switch and a safety interlock. Medical checks and specific training are required before installation or maintenance is carried out.

2.8. Summary

This chapter provided a review of FSO technology. Key features required of FSO technology to be suitable for use within the access network were highlighted, while the challenges posed by the atmospheric channel to an optical beam going through it were also discussed. Areas, where FSO can act as a data bridge, were mentioned, and lasers classification and safety requirements were also introduced.

3. THE ATMOSPHERIC TURBULENCE MODEL

3.1. Introduction

In order to predict the reliability of an optical system operating in different environments, it is very important to understand the statistical distribution of the received irradiance traveling through an atmospheric channel. In clear weather conditions, apart from attenuation, atmospheric turbulence also affects the FSO link performance [94, 95, 163]. Solar radiation absorbed by the Earth's surface induces warmer air around the Earth's surface than that at higher altitudes. Therefore, the warmer air being lighter rises to mix turbulently with the surrounding cooler air, thus causing the air temperature to fluctuate randomly [32, 96].

Inhomogeneities produced by turbulence can be regarded as discrete cells, or eddies of different temperature, behaving like refractive prisms of different sizes and indices of refraction [2, 50]. The interaction between the propagating optical beam (coherent focused laser beams) and the turbulent medium lead to random phase and amplitude variations (scintillation) of the information-bearing optical beam, thus resulting in fading (sometimes deep fading) and phase wondering (i.e. beam spreading) that degrade the performance of FSO links [97]. A number of mathematical models for the random fading irradiance distributions have been developed. However, due to the extreme complexity involved in mathematically modelling atmospheric turbulence, a valid and universal probability density function (pdf) that describes all turbulence regimes does not currently exist [1, 98]. The main reported model for irradiance fluctuations, namely the lognormal model corresponding to the weak regime [36, 63, 85, 94, 98], is elaborated upon and discussed in this chapter.

3.2. Optical Turbulence

Atmospheric turbulence is an outcome of the fluctuation of the atmospheric refractive index n along the path of the optical field/radiation passing through the atmosphere. This refractive index fluctuation is the direct and final result of the random variations in atmospheric temperature from point to point [32, 122]. Furthermore, these random temperature variations are a function of atmospheric pressure, altitude, and wind speed. The smallest and the largest of the turbulence eddies are named the inner scale, I_s, and the outer scale, O_s, of turbulence respectively. In general, I_s is in the order of a few millimetres, whereas O_s is typically in the order of several metres [29, 100]. These small sized prisms acting like eddies trigger a randomised interference effect between different points of the propagating beam, leading the wavefront to be distorted in the process, as shown in Figure 3-1.

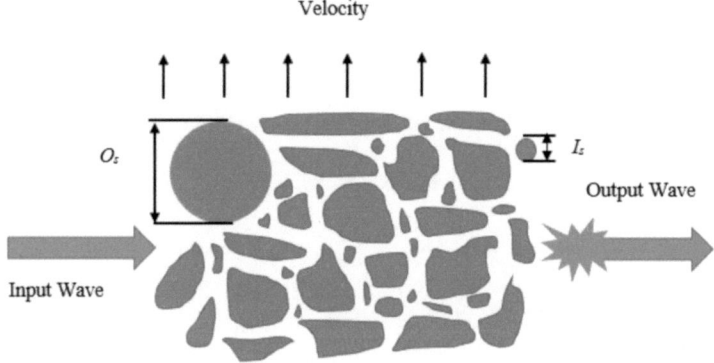

Figure 3-1: An atmospheric channel with velocity fluctuations and turbulent eddies.

Usually, O_s grow linearly within the height of the observation point over the ground in the surface layer to values close to 100 m. On the other hand, I_s near the ground is normally between 3 to 10 mm and can stretch to several centimetres in metres high altitudes $\hbar$ [100, 101].

The relationship between the temperature of the atmosphere and its refractive index is given by [102]:

$$n = 1 + 77.6(1 + 7.52 \times 10^{-3} \lambda^{-2}) \frac{P}{T_e} \times 10^{-6}, \tag{3.1}$$

$$-\frac{d_n}{d_{T_e}} = 7.8 \times 10^{-5} \frac{P}{T_e^2}, \tag{3.2}$$

where P is the atmospheric pressure in millibars. Humidity effects on the refractive index are considered negligible at specific bands [98].

Rytov variance σ_I^2 for a plane wave is defined as the turbulence strength measurement [1, 103, 104]:

$$\sigma_l^2 = 1.23 C_n^2 K^{7/6} L^{11/6}, \tag{3.3}$$

where C_n^2 refers to the index of refraction structure parameter, $K = 2\pi/\lambda$ is the optical wave number, and L(m) is the link length between T_x and R_x.

The value of C_n^2 changes with the altitude and is described by the Hufnagel-Valley (H-V) model, which is given by [1, 43, 70, 100]:

$$C_n^2(\hbar) = 0.00594(\upsilon/27)^2 (10^{-5}\hbar)^{10} \exp(-\hbar/1000) + 2.7 \times 10^{-16}$$
$$\exp(-\hbar/1500) + \hat{A}\exp(-\hbar/100), \tag{3.4}$$

where $\hat{A}$ is the normal value taken as the nominal value of C_n^2 (0) at the ground in $m^{-2/3}$, and v is the root mean square (RMS) wind speed (pseudo wind) in metres per second (m/s). The value of the index of refraction structure parameter changes with altitude, but for a horizontally spreading field, it is commonly assumed constant. In general, C_n^2 is defined as $10^{-12} m^{-2/3}$ for strong turbulence and $10^{-17} m^{-2/3}$ for weak turbulence. A typical average value is thus $10^{-15} m^{-2/3}$ [105].

In the spectral domain, the PSD of the refractive index fluctuation is closely related to C_n^2 via [106, 107]:

$$\Phi_n(K) = 0.033 C_n^2 K^{-11/3}, \quad 2\pi/O_s \ll K \ll 2\pi/I_s. \tag{3.5}$$

However, for a broad range of K, a different expression is derived by Tatarski and von Karman and described in [105, 108, 109]. Evaluating and mathematically characterising the turbulence is very challenging and complex due to the fact that observable atmospheric quantities are mixed non-linearly [100]. Hence, in order to deduce expressions for the statistical properties (namely the pdf and variance) of an optical beam spreading through the turbulent atmosphere, the following assumptions are employed to reduce mathematical details [105, 110]:

1) The atmospheric FSO channel is non-dispersive for wave propagation. This assumption can be explained by the fact that in the case of radiation and absorption of the optical beam by the atmosphere, the heat generated is unimportant compared to diurnal contributions.

2) The optical beam is not affected by any energy loss caused by atmospheric scattering. Consequently, the mean energy in the absence or presence of turbulence is the same. This assumption is valid for spherical and plane waves. Generally, a laser beam propagating over a long link span is considered to be a plane wave [98].

3.3. The Lognormal Turbulence Model

In reporting the pdf of the irradiance fluctuation in a turbulent atmosphere, the beam is first represented by its component electric field by employing Maxwell's electro-magnetic equations for the case of a spatially variant dielectric such as the atmosphere [45]:

$$\nabla^2 \vec{E} + k^2 n^2 \vec{E} + 2\nabla[\vec{E}.\vec{\nabla} ln(n)] = 0, \qquad (3.6)$$

where the vector gradient operator $\vec{\nabla} = (\frac{\partial}{\partial x})i + (\frac{\partial}{\partial y})j + (\frac{\partial}{\partial z})k$, with i, j and k being the unit vectors along the x, y and z-axes respectively. The last term on the left-hand side of (3.6) represents the turbulence induced depolarisation of the wave. In the case of weak atmospheric turbulence, which is characterised by a single scattering event, the wave depolarisation is negligible [36, 111, 112]. In fact, it has been demonstrated, both theoretically [112] and experimentally [113], that depolarisation is not important even for strong turbulence. Therefore (3.6) can be written as:

$$\nabla^2 \vec{E} + k^2 n^2 \vec{E} = 0 \tag{3.7}$$

The position vector is henceforward referred to by r, and $\vec{E}$ is represented by $E(r)$ for convenience.

In solving (3.7), Tatarski [108] introduced a Gaussian complex variable $\Psi(r)$, defined as the natural logarithm of the propagating field $E(r)$ and termed the Rytov transformation, which is given by:

$$\psi(r) = \ln[E(r)] \tag{3.8}$$

Rytov approach is also grounded on a fundamental assumption that the turbulence is weak and is characterised by a single scattering process. Evoking Rytov transformation (3.8), and matching the mean refractive index of the channel n_0 to unity, (3.7) transforms to the following Riccati equation, whose solution already exists

$$\nabla^2 \psi + (\nabla \psi)^2 + k^2 (1 + n_1)^2 = 0, \tag{3.9}$$

where n_1 represents the turbulence induced random fluctuation component. By means of the smooth perturbing method [95], the Gaussian complex variable $\Psi(r) = \ln\left[\overrightarrow{E_r}\right]$ can be written as [108]:

$$\psi(r) = \psi_0(r) + \psi_1(r), \tag{3.10}$$

where ψ_0 and ψ_1 represent the absence-of-turbulence part and the turbulence induced deviation respectively. Combining (3.10) with the Rytov change of variable (3.8) leads to the following:

$$\psi_1(r) = \psi(r) - \psi_0(r), \qquad (3.11)$$

$$\psi_1(r) = \ln[E(r)] - \ln[E_0(r)] = \ln[\frac{E(r)}{E_0(r)}], \qquad (3.12)$$

where the electric field in free space (with no turbulence) $E_0(r)$ is given by:

$$E(r) = A(r)\exp(i\phi(r)), \qquad (3.13a)$$

$$E_0(r) = A_0(r)\exp(i\phi_0(r)), \qquad (3.13b)$$

where $A(r)$ and $\phi(r)$ represent the amplitude and phase of the field with atmospheric turbulence respectively, whilst $A_0(r)$ and $\phi_0(r)$ stand for the amplitude and phase of the field with no atmospheric turbulence respectively. Therefore, (3.12) can be rewritten as [1, 114]:

$$\psi_1(r) = \ln[\frac{A(r)}{A_0(r)}] + i[\phi(r) - \phi_0(r)] = X + i\delta, \qquad (3.14)$$

As $\psi_1(r)$ is Gaussian, X refers to the Gaussian distributed log-amplitude fluctuation and δ represents the Gaussian distributed phase fluctuation of the field. By focusing solely on the field amplitude, however, the pdf of X can be expressed by [45, 47]:

$$P(X) = \frac{1}{\sqrt{2\pi\sigma_x^2}} \exp\left\{-\frac{(X-E[X]^2)}{2\sigma_x^2}\right\}, \qquad (3.15)$$

where $E[X]$ represents the expectation of X and σ_x^2 denotes the log-amplitude variance, usually referred to as the Rytov parameter. According to [30], σ_x^2 is what qualifies the extent of field amplitude fluctuation in atmospheric turbulence and is associated with C_n^2 and the horizontal distance, L, and given by:

$$\sigma_x^2 = 0.56 k^{7/6} \int_0^L C_n^2 (L-x)^{5/6} dx \quad \text{for a plane wave,} \qquad (3.16)$$

and

$$\sigma_x^2 = 0.56 k^{7/6} \int_0^L C_n^2 (x/L)^{5/6} (L-x)^{5/6} dx \quad \text{for a spherical wave} \qquad (3.17)$$

For a field spreading horizontally over a turbulent medium, as is the case in major terrestrial applications, C_n^2 remains constant, and the log irradiance variance for a plane wave turns into:

$$\sigma_I^2 = 1.23 C_n^2 k^{7/6} L^{11/6} \qquad (3.18)$$

The field irradiance (intensity) in the turbulent medium is $I = |A(r)|^2$, while the intensity of free space without turbulence is $I_0 = |A_0(r)|^2$. The log-intensity is therefore given by:

$$l = \log_e \left| \frac{A(r)}{A_0(r)} \right|^2 = 2X. \qquad (3.19)$$

Therefore,

$$I = I_0 \exp(l) \qquad (3.20)$$

In order to get the irradiance pdf, the variable transformation $P(I) = P(X)\left|\frac{dX}{dI}\right|$ is invoked, and log-normal pdf is obtained via [103, 104]:

$$P(I) = \frac{1}{\sqrt{2\pi\sigma_l^2}} \frac{1}{I} \exp\left\{-\frac{(\ln(I/I_0) - E[l])^2}{2\sigma_l^2}\right\} \quad I \geq 0. \qquad (3.21)$$

In regions of weak turbulence, the statistics of the irradiance fluctuations, reported experimentally in [49], indicate that they obey log-normal distribution.

From (3.19), the log-intensity variance $\sigma_l^2 = 4\sigma_x^2$ and the mean log intensity $E[l] = 2E[X]$. Based on the second assumption in Section 3.2, $E[\exp(l)] = E[I/I_0] = 1$. The amount of energy is considered to be the same during the scattering process, which means $E[I] = I_0$. Using the standard relation (3.22) (which is valid for a Gaussian random variable with real values) [105], the expression for $E[l]$ is given by [164]:

$$E[\exp(az)] = \exp(aE[z] + 0.5a^2\sigma_z^2), \qquad (3.22)$$

$$1 = \exp(E[l] + 0.5\sigma_l^2), \qquad (3.23)$$

$$E[l] = -\sigma_l^2/2, \qquad (3.24)$$

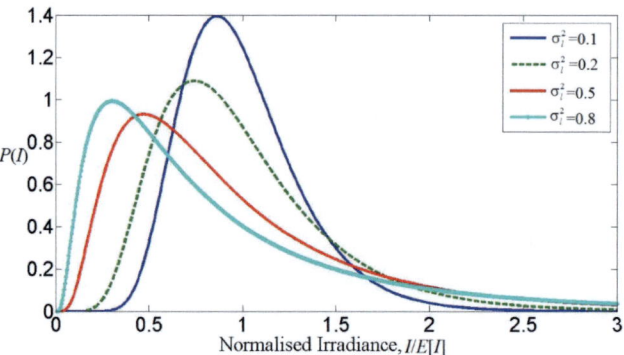

Figure 3-2: Normalised log-normal pdf for a range of irradiance variance σ_I^2 values.

After obtaining the pdf of the irradiance fluctuation, it is also possible to derive an expression for the variance of the irradiance fluctuation σ_I^2, which characterises the strength of irradiance fluctuation. The log-normal pdf is plotted in Figure 3-2 for various values of log irradiance variance. The normalised variance of the irradiance fluctuation, also referred to as the scintillation index (S.I), which mainly characterises the effects of atmospheric turbulence, is hence given by [100]:

$$\text{S.I} = \sigma_N^2 = \frac{\sigma_I^2}{I_0^2} = \exp(\sigma_0^2) - 1, \qquad (3.25)$$

When S.I. increases linearly with σ_I^2 until it attains a maximum value greater than unity, the scintillation transitions into the strong turbulence regime qualified by random focusing due to the large-scale inhomogeneities [36, 98]. Continuously increasing the inhomogeneity strength causes the decrease of the focusing effect and the peak fluctuations as a direct outcome of multipole self-interferences. In this case, the turbulence moves into the saturation regime, at which the value of the scintillation index approaches unity as Rytov parameter increases [98].

3.4. Spatial Coherence in Weak Turbulence

Due to temperature, velocity and refractive index fluctuations, the spatial coherence of the beam degrades during propagation over the atmospheric channel. The extent of this coherence degradation is a function of the atmospheric turbulence strength and the propagation distance. As per Rytov approach, applied in the modelling of weak atmospheric turbulence, the spatial coherence of a field propagating over the atmosphere can be expressed by [36, 100, 111]:

$$\Gamma_X(\rho) = A^2 \exp[-(\rho/\rho_0)^{5/3}], \qquad (3.26)$$

Where ρ_0 refers to the transverse coherence length at which the coherence of the optical field is equal to e^{-1}.

The transverse distance ρ_0 for the plane and spherical waves is given by the following equations respectively [36]:

$$\rho_0 = [1.45 k^2 \int_0^L C_n^2(x) dx]^{-3/5}, \qquad (3.27)$$

$$\rho_0 = [1.45 k^2 \int_0^L C_n^2(x)(x/L)^{5/3} dx]^{-3/5}. \qquad (3.28)$$

The coherence length is useful in determining the size of the R_x aperture. The coherence length reduces as the link range or the turbulence level rises. A comparison of the spatial coherence length (in meters) for a plane wave in strong turbulence $C_n^2 = 10^{-12} m^{-2/3}$ nm and weak turbulence $C_n^2 = 10^{-15} m^{-2/3}$ at λ of 830 and 1550 nm are given in Tables 3-1 and 3-2 respectively.

Table 3-1: The dependence of the spatial coherence length on wavelength and link distance for $C_n^2 = 10^{-12} m^{-2/3}$.

Link distance (km)	0.1	1	10	100
Spatial coherence length (m) at λ of 830 nm	2.8	0.7	0.18	0.04
Spatial coherence length (m) at λ of 1550 nm	6.5	1.6	0.3	0.09

Table 3-2: The dependence of the spatial coherence length on wavelength and link distance for $C_n^2 = 10^{-15} m^{-2/3}$.

Link distance (km)	0.1	1	10	100
Spatial coherence length (m) at λ of 830 nm	0.5	0.1	0.05	0.01
Spatial coherence length (m) at λ of 1550 nm	0.7	0.2	0.09	0.04

3.5. Summary

In this chapter, characterisation of different atmospheric turbulence models was presented, covering short to very long range FSO links. The log-normal distribution was discussed and its limitations highlighted. This model is mathematically presented as tractable, however, it is only valid for the weak turbulence regime. In other cases where multiple scattering needs to be accounted for (i.e. a strongly turbulent atmosphere), the Gamma-Gamma model is more suitable but lacks the mathematical convenience of the log-normal distribution. The log-normal model will be applied in the subsequent chapters to characterise the statistical behaviour of the received signal, and derive expressions for the error performance of short to very long FSO links.

4. FSO MODULATION TECHNIQUES

4.1. Introduction

This chapter introduces the modulation techniques widely employed in FSO, and discusses their performance through channel impairments such as noise and channel fading caused by atmospheric turbulence. The selection of the most appropriate modulation scheme in the design process an FSO system is very crucial for the performance of the communication system [116, 120, 125, 126]. Prior to the selection of a modulation scheme, it is sensible to define the criteria of adopting a modulation scheme in communication systems [127, 128].

There are many different types of modulation schemes which are suitable for FSO communications. Similar to other communication systems, when choosing a modulation scheme for FSO systems, three important metrics need to be considered [43]:

1) Power efficiency: Due to skin and eye safety considerations, the emitted optical power permitted is limited. Therefore, modulation schemes with a high peak and low average power would be the most suitable options.

2) Bandwidth efficiency: This may be more important for non-FSO systems.

3) System equipment: In FSO systems, the optical carrier can be said to have 'unlimited bandwidth', but the other constituents of the systems (e.g. photodetector) limit the bandwidth practically available.

4) Simplicity: The simplicity of a modulation scheme is very important as directly affects the cost of implementation.

This chapter focuses on the following digital modulation techniques: OOK, BPSK-SIM, PPM, QPSK and hybrid BPSK-SIM-PPM. Due to the fact that the average emitted optical power is always limited, we compared the performance of the modulation techniques in terms of the average received optical power to attain a desired bit error rate at a given data rate. While it is very important for the modulation scheme to be power efficient, this is not the only determining factor in the selection of a modulation technique. The design complexity of the corresponding T_x and R_x and the bandwidth requirement of the modulation scheme are all equally significant.

OOK is the classical often-used modulation technique [27, 53]. This is principally due to its design and implementation simplicity, which is the reason why the major reported research studies in the literature [49, 51] are established on this signalling technique. However, the performance of fixed threshold level OOK in atmospheric turbulence is not optimal, as will be highlighted in the following section. In atmospheric turbulence, an optimal performing OOK necessitates the threshold level to change according to the persisting irradiance fluctuation and noise (that is, to be adaptive) [1]. The PPM scheme expects no adaptive threshold and is predominantly employed for deep space FSO communication links due to its increased power efficiency compared to the OOK scheme [129, 130, 131]. On the other hand, the PPM modulation technique requires a complex transceiver design due to tight synchronisation requirements and a higher bandwidth than OOK.

The performance analysis of these modulation schemes and that of the SIM and QPSK schemes are further discussed in this chapter. The SIM scheme also adopts no adaptive threshold, and does not require as much bandwidth as PPM, but suffers from a high peak to average power ratio, which translates into poor power efficiency. Opting for a modulation scheme for a specific application hence implies trade-offs among these highlighted factors.

4.2. On-Off Keying

OOK is the prevailing modulation scheme used in a commercial terrestrial wireless optical communications system; this is mainly because of its simplicity and resilience to the nonlinearities of the laser and the external modulator. OOK can employ either non-return-to-zero (NRZ) or return-to-zero (RZ) pulse formats. In NRZ-OOK an optical pulse of peak power $\alpha_e P_T$ stands for a digital symbol '0' whilst the transmission of an optical pulse of peak power presents a digital symbol '1'. The optical source extinction ratio, α_e, has the range $0 < \alpha_e < 1$. The finite duration of the optical pulse is the same as the symbol duration T_{ps}. In the case of OOK-RZ T_{ps} is lower than the bit duration T_b, which results in amelioration in power efficiency over NRZ-OOK at the expense of a raised bandwidth requirement. In all the analyses that follow the α_e, is equal to zero, and NRZ-OOK, which is the common scheme used in current commercial FSO systems, is presumed unless otherwise submitted.

4.2.1. OOK-NRZ

In NRZ-OOK, an optical pulse represents a digital symbol '1', while the transmission of no optical pulse represents a digital symbol '0'. In order to simplify the modulator, the pulse shape is generally chosen to be rectangular [132]. Adapting T_b, the bit rate is then expressed as:

$$R_b = 1/T_b, \qquad (4.1)$$

The normalised transmit pulse shape for OOK is defined by:

$$g(t) = \begin{cases} 1 & \text{for } t \in [0, T_b) \\ 0 & \text{otherwise} \end{cases}, \qquad (4.2)$$

In the demodulator, the received pulse is integrated over the one-bit period, and subsequently sampled and compared to a threshold to decide a '1' or a '0' bit. This is acknowledged as the maximum likelihood R_x, which minimises the BER [62].

Moreover, an important parameter that has to be taken into consideration in any modulation scheme is the bandwidth requirement. The bandwidth is estimated by the first zeros in the spectral density of the signal. The spectral density is calculated via the Fourier transform (FT) of the autocorrelation function. The electrical power density for OOK-NRZ considering the independently and identical distributed (IID) input bits is provided by [62, 133]:

$$S_{OOK-NRZ}(f) = (P_r R)^2 T_b \left(\frac{\sin \pi f T_b}{\pi f T_b} \right) \left[1 + \frac{1}{T_b} \delta(f) \right]. \qquad (4.1)$$

where P_r is the average electrical power and R stands for the response of the R_x. The OOK-NRZ modulation format changes in pulse width; that is the pulse shape is high for only a fraction of bit duration δT_b, with $0 \le \delta < 1$. The advantage of this scheme is to attain a reduction in the transmitted power. Nevertheless, as δ decreases, the bandwidth requirement increases faster than the diminution in power requirement. Therefore, this type of OOK is inferior to PPM, which requires less bandwidth to achieve a given reduction in power. For δ = 0.5, this scheme is usually named OOK-RZ [96, 62]. In OOK-RZ, the power requirement is reduced to half of the regular OOK-NRZ, at the expense of doubling the bandwidth. The formula for the electrical power density for OOK-RZ, considering random input bits, is as follows [133]:

$$S_{OOK-RZ}(f) = (P_r R)^2 T_b \left(\frac{\sin \pi f T_b / 2}{\pi f T_b / 2} \right)^2 \left[1 + \frac{1}{T_b} \sum_{-\infty}^{\infty} \delta \left(f - \frac{n}{T_b} \right) \right]. \quad (4.4)$$

Figure 4-1 depicts the power spectrum for the OOK-NRZ and OOK-RZ (δ= 0.5) schemes. The power axis is normalised to the average electrical power P_r multiplied by T_b and the frequency axis is normalised to R_b.

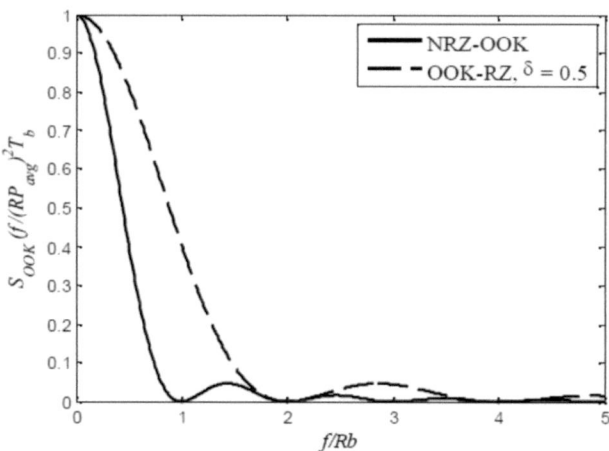

Figure 4-1: The power spectrum of the transmitted signals for OOK-NRZ and RZ [62].

4.3. Pulse Position Modulation

In FSO communications, it is crucial to adopt power efficient modulation schemes due to the fact that the requirement for bandwidth is not a major issue. This is because most of FSO links are LOS links employing a laser with abundant bandwidth. The PPM modulation scheme offers more power efficiency than OOK, but at the expense of an increased bandwidth requirement and higher complexity [134]. In L-PPM each word of M bits is mapped into one of $L=2^M$ symbols and transmitted via the available channel. An L-PPM symbol has the form of a pulse transmitted in one of $L=2^M$ consecutive time slots with duration $T_s=MT_b/L$, with the remaining slots being empty (see Figure 4-2). Information is encoded within the position of the one pulse of constant power along with (M-1) empty slots. The position of the pulse matches the decimal values of the M-bit input data [135].

The transmit pulse shape for L-PPM is defined by [124]:

$$x(t)_{PPM} = \begin{cases} 1 & for\ t \in \left[(m-1)T_{s\_PPM}, T_{s\_PPM,mT}\right] \\ 0 & otherwise \end{cases}, \quad (4.5)$$

where $m \in \{1,2,3......L\}$.

Therefore, the sequence of the PPM symbol can be expressed by:

$$x(t)_{PPM} = LP_r \sum_{k=0}^{L-1} C_k g\left(t - \frac{KT_{symb}}{L}\right), \quad (4.6)$$

where $C_k \in \{C_0, C_1, C_2....C_{L-1}\}$ denotes the PPM symbol sequence, $g(t)$ is the pulse shape function of unity height having a duration of T_{symb}/L, T_{symb} ($=T_b M$) is the symbol interval and LP_{avg} is the peak optical power of the PPM symbol [124].

In L-PPM, all signals are equidistant, with:

$$d_{\min-PPM} = \min_{i \neq j} \int \left[x_i(t) - x_j(t) \right]^2 dt = 2LP^2 \log_2(\frac{L_{slot}}{R_b}), \qquad (4.7)$$

where L_{slot} is the slot rate and P is the average power. The transmitted waveforms for 16-PPM and OOK are shown in Figure 4-2. Detection of the L-PPM symbols requires the estimation of the slot where the pulse was most probably transmitted. Nevertheless, because of its superior power efficiency, PPM is an attractive modulation scheme for FSO communications and deep space laser communication applications [136]. The infrared physical layer of IEEE 802.11 standard on wireless local area networks (LANs) recommends 16-PPM for 1Mbps data rates and 4-PPM for 2 Mbps data rates [124].

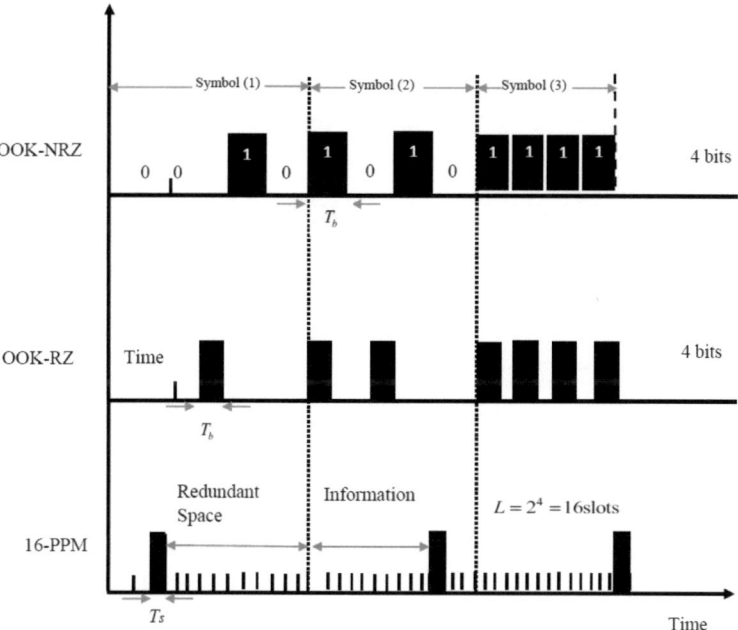

Figure 4-2: Time domain waveforms for 4-bit OOK and 16-PPM.

Due to the fact that the average emitted optical power is mostly limited, the performance of modulation techniques is usually compared in terms of the average received optical power required to achieve the desired BER at a defined data rate [1, 137]. Figure 4-3 depicts the SNR required to attain a particular BER for OOK-NRZ, OOK-RZ and 4-PPM. 4-PPM requires almost 6 dB less power to achieve a BER of 10^{-9} compared to OOK-NRZ.

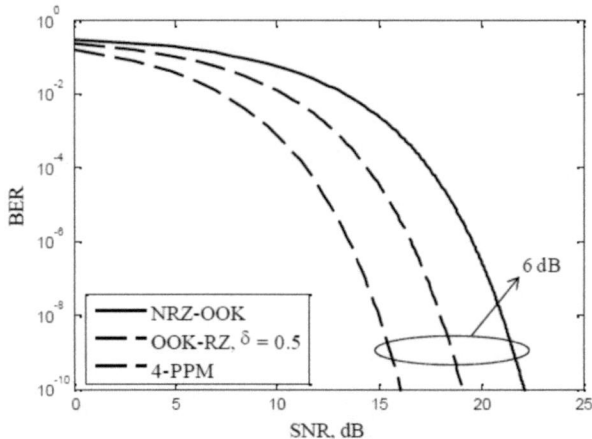

Figure 4-3: BER performance for OOK (NRZ and RZ) and 4-PPM [49].

4.4. Subcarrier Intensity Modulation

SIM is a technique adopted from the very successful multiple carrier RF communications already used in applications such as LANs, asymmetric digital subscriber line (ADSL), digital television, 4G communication systems, and optical fibre communications [138, 139]. For instance, in optical fibre communication networks, the subcarrier modulation techniques have been commercially adopted in carrying cable television signals, and applied in conjunction with WDM [140]. For the smooth integration of FSO systems into present and future networks that already hold subcarrier modulated (or multiple carriers) signals, the study of subcarrier modulated FSO is hence important. There are several reasons for considering subcarrier intensity modulated FSO systems, including:

a) Benefiting from the already developed and established RF communication components, such as stable oscillators and narrow filters [141].

b) Avoiding the requirement for an adaptive threshold needed by optimum performing OOK-modulated FSO [1, 117].

c) Can be employed to increase capacity by adapting data from different users on different subcarriers.

d) Such systems have a relatively lower bandwidth requirement than PPM-based systems.

On the other hand, there are some challenges facing the deployment of SIM systems, including:

a) Relatively high average transmitted power due to:

1) The optical source being ON throughout the transmission of both binary digits '1' and '0', which different from OOK where the source is just ON during the transmission of bit '1'.

2) The multiple subcarrier composite electrical signal, being the sum of the modulated sinusoids (i.e. dealing with both negative and positive values), requires a direct current (DC) bias. This is to ensure that this composite electrical signal, which will eventually modulate the laser irradiance, is never negative. Enhancing the number of subcarriers N results in increased average transmitted power since the minimum value of the composite electrical signal step-downs (becomes more negative) and the required DC

bias consequently increases [111]. This factor results in poor power efficiency and places a limit on the number of subcarriers that can be admitted when using multiple SIM.

b) The possibility of signal deformations due to the inherent laser non-linearity and signal clipping caused by over-modulation.

c) The rigorous synchronisation requirements at the R_x side.

In BPSK-SIM, the RF subcarrier pre-modulated with data $d(t)$ is applied to modulate the intensity of the optical carrier. Figure 4-4 describes the modulation and demodulation processes in a BPSK-SIM FSO link. Prior to modulating, the optical carrier $d(t)$ is modulated by the RF subcarrier BPSK signal, where bits '1' and '0' are corresponded with a phase shift of 180°. Unlike baseband modulation schemes (OOK, PPM and pulse amplitude modulation (PAM)) in which information is encoded in the amplitude of the carrier [49]), the information in BPSK-SIM is encoded in the phase of the RF carrier. This grants higher immunity to intensity fluctuation and removes the need to adaptive thresholding [1, 143].

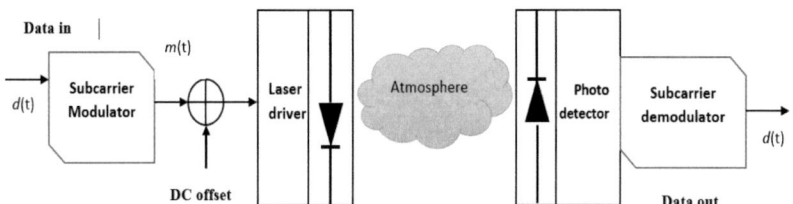

Figure 4-4: Block diagram of a BPSK subcarrier intensity-modulated FSO link.

The received photocurrent can be modelled as [124]:

$$i_r(t) = d_j RI \xi A g(t) \cos(w_c t) + n(t), \qquad (4.8)$$

where $d_j \in \{1,-1\}$ is the signal level for the *j*th data symbol matching to the data symbol '1' and '0', $I=0.5I_{peak}$, where I_{peak} is the obtained peak irradiance, *R* is the photodetector responsivity, *g(t)* is the pulse shape function, $n(t) \approx N(0,\sigma^2)$ is the additive white Gaussian noise (AWGN), which is an outcome of the thermal and the background noise, and ξ is the optical modulation index.

By reducing the requirement for adaptive thresholding, as it is held at the zero mark [124] resulting in BPSK being simpler than OOK in an atmospheric turbulence channel, and considering the AWGN channel and the post-detection electrical signal-to-ratio ratio (SNR$_e$) at the input of the BPSK, the demodulation process is expressed by [1]:

$$SNR_e = \frac{(IR\xi)^2 P_m}{2B_e(qRI_{Bg} + 2k_BT_e/R_L)}, \qquad (4.9)$$

where $P_m = (1/T)\int_T m^2(t)dt$ is the subcarrier signal power, *Be* is the post detection electrical filter bandwidth expected to pass the information signal *m(t)* without distortion, and I_{Bg} stands for the background radiation irradiance. Multiple-SIM could be applied for higher capacity/users at the cost of poor power efficiency [144, 145]. In this book, the previously mentioned modulation schemes are experimentally implemented and compared as a means of battling the effect of atmospheric turbulence and fog in Chapters 6 and 7.

4.5. Quadrature Phase Shift Keying (QPSK)

QPSK is a commonly used modulation technique in fibre optical coherent communications. QPSK can be referred to as a method for transmitting digital information across an analogue channel. Data bits are grouped into pairs, and each pair is represented by a particular waveform, called a symbol, to be sent across the channel after modulating the carrier [146]. The QPSK modulation technique demonstrates a good compromise between R_x complexity, bit error rate, data rate, and bandwidth for free space links. It has twice the data rate for a given bandwidth compared to BPSK while holding the same bit error rate using heterodyne detection, leading to better performance in terms of sensitivity than other coherent systems (such as Amplitude Shift Keying (ASK) and Frequency-shift keying (FSK)), and permitting the implementation of higher modulation orders without increasing the complexity of the system [147, 148].

The fundamental function of a QPSK modulation scheme can be expressed by [148]:

$$\emptyset_0(t) = \sqrt{2}p(t)\cos[w_c t], \qquad (4.10)$$

$$\emptyset_1(t) = \sqrt{2}p(t)\sin[w_c t], \qquad (4.11)$$

where the intermediate RF frequency $w_c = 2\pi f_c$, f_c is the carrier frequency and $p(t)$ is a unit energy pulse shape of T_{ps} duration.

This is an orthonormal basis of the sinusoid, and both channels are autonomous from one another. The resulting signal comprising the information can be expressed as follows:

$$s(t) = \sqrt{2}\sum_k x_0(k)p(t - kT_{samp})\cos[w_c t] - x_1(k)p(t - kT_{samp})\sin[w_c t], \qquad (4.12)$$

where $x_0(k)$ and $x_1(k)$ are the *k-th* transmitted symbol and T_{samp} is the sampling time. This expression can be written as:

$$s(t) = I(t)\sqrt{2}x_0(k)p(t-kT_{samp})p(t-kT_{samp}) + Q(t)\sqrt{2}x_1(k)p(t-kT_{samp}) \quad (4.13)$$

where $I(t)$ refers to the in-phase channel (I) and $Q(t)$ stands for the quadrature channel (Q), and are defined by:

$$I(t) = \sum_k x_0(k)p(t-kT_{samp}), \quad (4.14)$$

$$Q(t) = \sum_k x_1(k)p(t-kT_{samp}) \quad (4.15)$$

The symbol constellation for QPSK modulation with Gray coding is shown in Figure 4-5. The relationship between the two signals, I and Q, define the demodulated data bits.

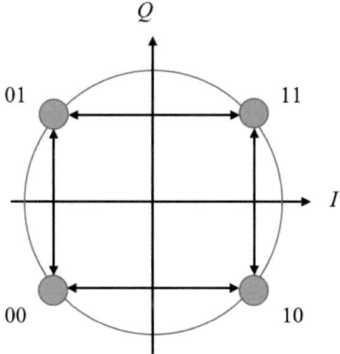

Figure 4-5: QPSK symbol constellation with Gray coding.

In order to build up a QPSK communication system, first, a pseudo-random sequence of b data bits is generated. It is important to mention that b is a bit array of d_b length, and it plays the role of transmitting the information message. After that, a unique word preface w of length d_w is added before b. The sum of both lengths is equal to $2k$. This preamble is required in order to find the beginning of the message and to design an ambiguity resolution circuitry at the R_x that will support the demodulation process

[148, 149]. The data bits are transformed from serial to parallel in order to form pairs of bits. Each group of pairs generates a pair of signals by employing a lookup table comprising the symbol constellation. In this block, $x_0(k)$ and $x_1(k)$ are generated. The transmission filter is then used by multiplying each signal by $p(t)$, generating $I(t)$ and $Q(t)$ from equations 4.14 and 4.15 respectively. The up-conversion process is then implemented by multiplying both signals by $\cos[W_c t]$ and $\sin[W_c t]$ respectively, and the outcome signals are added to generate $s(t)$, as illustrated in Figure 4-6.

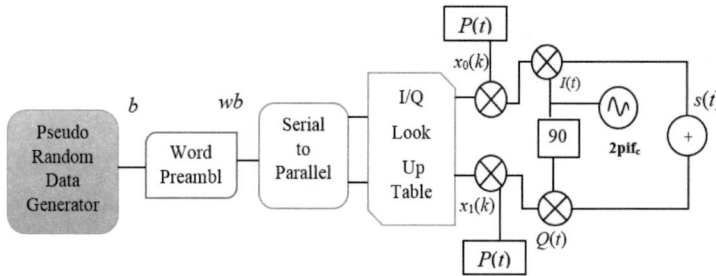

Figure 4-6: Heterodyne T_x block system [148].

4.6. Summary

This chapter presented a detailed literature review of the FSO communication modulation systems. The selection of modulation technique requires a balance between power efficiency, simplicity and bandwidth efficiency. Considering the BER performance of OOK, the poor bandwidth efficiency of PPM, and the required slot and symbol synchronisation requirements, the focus of this work will be on the SIM (which nevertheless suffers from poor power efficiency due to its high peak to average power ratio) and the QPSK modulation techniques. In the next chapter, the design of a controlled atmospheric channel will be presented.

5. DESIGN OF THE ATMOSPHERIC CHANNEL

5.1. Introduction

The real outdoor atmosphere (ROA) channel inflicts a number of challenges on FSO links [149, 150]. The ROA is time varying and depends on the magnitude and intensity of weather conditions [151, 152]. These conditions have various effects on the FSO link performance, with dense fog and intense scintillation inducing the most severe deterioration to the FSO link performance [118, 153, 154]. Fog, smoke, aerosols, rain, snow, dust and turbulence conditions, as well as the mixture of some of these phenomena, forms the ROA. Consequently, carrying out a proper link assessment under particular weather conditions is a challenging task [155, 156]. Investigating the real outdoor fog (ROF), which is well known as heterogeneous and very unpredictable in nature, is an important challenge in FSO communication links [49, 155]. This is due to various reasons such as: (i) the inaccessibility of an experimental setup for outdoor links due to the long observation time and the low reoccurrence of fog (e.g. dense fog events for visibility $V < 0.5$ km), (ii) the difficulty in controlling and characterising the atmosphere due to the presence of aerosols such as fog and smoke, which are inhomogeneous, along the FSO link path, (iii) the expensive, and not promptly available outdoor measurement instruments, and (iv) the repetition of the same measurement condition several times when needed is not practicable.

Moreover, atmospheric turbulence (scintillation) has been widely investigated, and a number of theoretical models have been proposed to model scintillation induced fading [157]. Nevertheless, it is very challenging to practically measure the impact of atmospheric turbulence under various weather conditions, primarily due to (i) the long waiting time required to observe and experience the reoccurrence of atmospheric turbulence events, which usually can take a very long time, and (ii) the difficulty in controlling and characterising the atmospheric turbulence. Therefore, in this research, an indoor atmospheric laboratory chamber was designed so that the ROF, smoke and the atmospheric turbulence channel can be simulated under controlled conditions. Hence, particular measurements to investigate the effects of smoke, the ROF, and turbulence on the link performance can be carried out. The laboratory chamber provides the advantage of a full system characterisation and assessment in a shorter time compared with the outdoor FSO link, where it often takes a long time for the weather conditions to change.

In this chapter, a detailed description of the design of the laboratory atmospheric chamber is introduced. Furthermore, the methods to produce turbulence and homogeneous fog are discussed, and the data acquisition techniques for measuring both the turbulence- and fog-induced attenuation are presented. The methods used to produce and control atmospheric turbulence are also discussed.

5.2. The Atmospheric Chamber

A snapshot of the laboratory-based testbed consisting of the atmospheric chamber used to control and simulate the ROA conditions is shown in Figure 5-1. The block diagram of the chamber is discussed in detail in [158]. The laboratory atmospheric channel is a closed glass chamber with multiple compartments (three in total for this experiment), and each has a vent to let air circulate into and out of the channel. The temperature and wind velocity conditions in the chamber are controlled to simulate the atmospheric conditions as closely as possible. By applying a number of heaters and fans, it is possible to generate and control the temperature induced turbulence inside the chamber. The controlled temperature, and the wind velocity inside the chamber can be promptly controlled to simulate the outdoor atmospheric conditions as closely as possible. Furthermore, this designed chamber permits carrying out the characterisation of the outdoor FSO system as well as the performance measurements under several levels of turbulence without having to wait for long observation times; which is the case in real weather conditions. Table 5-1 presents the key parameters of the designed chamber.

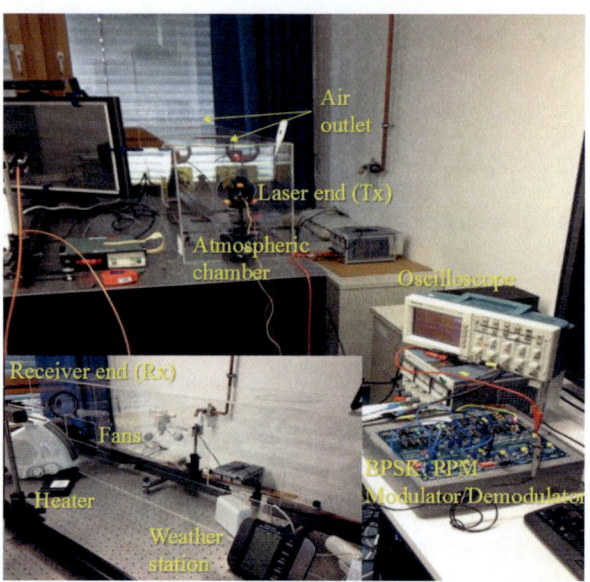

Figure 5-1: A snapshot of an atmospheric chamber in the lab.

Table 5-1: Main parameters of the designed lab-chamber.

Parameter	Value
Dimension	300×30×30 cm^3
Temperature range	20 - 80 °C
Wind speed	3-5 m/s
Heater machine	HL 3379 Fan heater, 1000 Watt
Fog machine	ADJ MINI FOG 400 Steamer

Within the atmospheric chamber, the specific effects of turbulence and fog are controlled in order to mimic the ROA as closely as possible. The fog is generated by using a commercial water vaporising machine (water steam) with 100% humidity to simulate the ROF [159].

5.3. The Intelligent Weather Station

Accurate measurements of chamber atmospheric temperature, relative humidity, pressure and turbulences as well as the availability of predictions of their evolution over time are important in order to simulate the ROA. Hence, it is of significant importance to use a very accurate, closed loop intelligent weather station (IWS). A lightweight and portable sensor-based intelligent weather station was employed, utilising nearest-neighbours (NEN) algorithm models as the time-series predictor mechanisms.

Using forecasts and a feedback loop of solar radiation, air temperatures, humidity and pressure reduces the risk of temporary depletion and increases data usability. As such, this illustrates the need of a weather station that, besides measurement of atmospheric variables, provides the forecasts autonomously and makes them available wirelessly to the application at hand. The IWS monitors and archives different chamber controlled parameters such as temperature, air pressure, and wind velocity, which are employed in order to perform measurements on the effects of scintillation on the FSO link, as well as to maintain the values for different measurements over a long period of time.

In the experimental work, the IWS Conrad Radio Digital Weather Station WS1600 was used. It is also equipped with the innovative Weather Center button, which shows additional data about the weather on a given day (i.e., storing up to 200 sets of weather data, which are recorded automatically at 3-hour intervals after the weather station is powered up) [215]. The IWS described in Figure 5-2 and Figure 5-3, includes two components: the Integrated Sensor Suite (ISS), which houses and manages the external sensor array, and the console, which provides the user interface, data display, and calculations. The IWS communicates via an FCC-certified, license ‑ free frequency-hopping transmitter and receiver. The frequency ‑ hopping spread ‑ spectrum (FHSS) technology provides greater communication strength over longer distances and in areas of weaker reception. User-selectable transmitter ID codes allow up to eight stations to coexist in the same geographic area. The WeatherLink® allows the weather station to interface with a computer, log weather data, and upload weather information to the internet [216]. The IWS is described in the block diagram in Figure 5-2. It measures several atmospheric variables such as the global solar radiation, air temperature, pressure and relative humidity.

These measurements are carried at a user-determined time interval. Moreover, the IWS predicts the development of each variable in a prediction horizon of up to 40 steps-ahead. The prediction step can coincide with the measurement sampling time, or can be a multiple of it. In this case, the average value of each variable is computed over the prediction step. In the case presented here, the maximum prediction horizon is 24 hours. The system was built with the objective of being self-sufficient regarding electrical energy. Furthermore, to facilitate its deployment, it integrates wireless communication based on the wireless IEEE 802.15.4 standard [215].

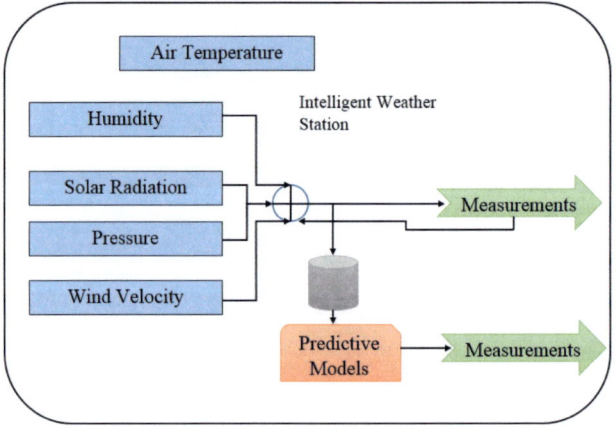

Figure 5-2: IWS signal-flow block diagram.

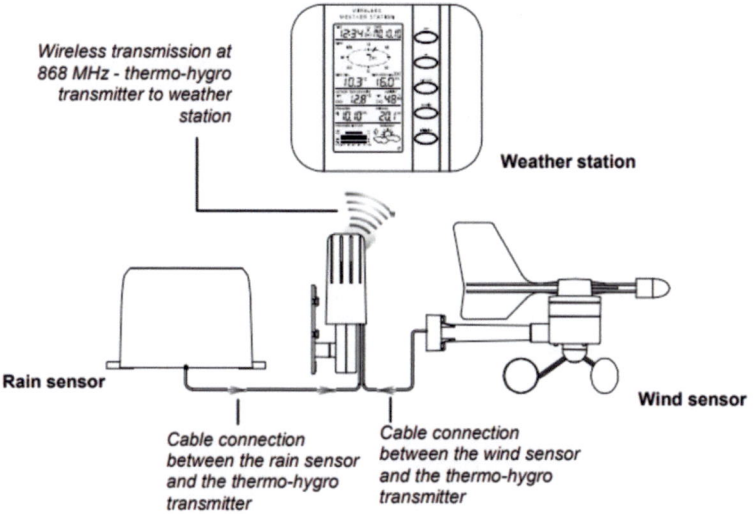

Figure 5-3: A snapshot of the Conrad Digital Radio Weather Station WS1600 [215].

5.4. The Controlled Turbulence Simulation Channel

A schematic diagram of the turbulence simulation chamber is presented in Figure 5-4. The chamber has dimensions of 300×30×30 cm. The scintillation effect induced by atmospheric turbulence is simulated by exploiting the dependence of the channel index of refraction on the temperature fluctuations, as earlier discussed in Chapter 3.

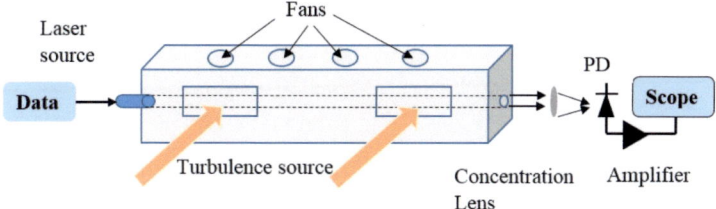

Figure 5-4: A block diagram of an FSO experimental system with turbulence simulation.

The turbulence simulation process requires blowing cold and hot air into the chamber at various locations, as depicted in Figure 5-4. The direction of air is such that it is orthogonal to the optical beam's direction of propagation. The cold air is set at a room temperature of ~ 27 °C and the hot air is set at a temperature range from 27 to 95 °C. In this chamber, there are two approaches that could be used to create the turbulence effect.

a) Using a heater and fans to blow hot and cold air in the direction perpendicular to the optical beam propagating along the chamber in order to generate a variation in temperature and wind speed, thus enabling the experimental investigation of the weak and medium turbulence, as given by the theoretical model in (3.3) (see Figure 5-2 and 5-3). The cold air is at room temperature (20 - 27° C) and the hot air temperature is in the range between 20 to 70° C. Using a series of air vents, extra temperature control is achieved, thus ensuring a constant temperature gradient between the source and the detector.

b) Each compartment has a powerful internal heating source inside the chamber and a fan attached to its vent so that a very strong turbulence effect can be generated. A similar approach for simulating turbulence has been reportedly used by other researchers [1, 49, 65, 121].

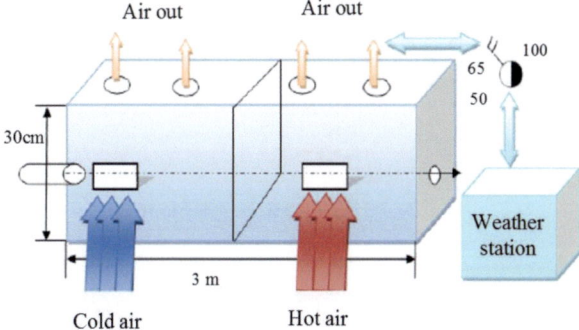

Figure 5-5: The laboratory turbulence chamber activated by hot and cool air.

In order to compensate for the short chamber length, $\Delta T / \Delta L \approx 5°K/m$ the temperature measured at each point along the chamber is retained within the tolerance margin of $\pm 1° K$. The average temperatures recorded at four different locations along the chamber are presented in Table 5-2.

Table 5-2: Measured temperatures over six experiments at four different positions within the chamber.

	T1(°C)	T2(°C)	T3(°C)	T4(°C)	$C_n^2(m^{-2/3})$
Set 1	29	24	22	21	1.60 x 10^{-12}
Set 2	35	30	25	23	5.61 x 10^{-12}
Set 3	36	29	24	22.8	9.50 x 10^{-12}
Set 4	42	33	28	26	3.04 x10^{-12}
Set 5	53	41.5	28	25	4.02 x 10^{-12}
Set 6	58	45	33	29	4.90 x 10^{-12}

Appling (5.1), the average values of C_n^2 against a range of temperature gradients within the chamber is also shown in Table 5-2. The average values of C_n^2 vary from 10^{-12} to 10^{-10}, which is in good agreement with the experimental data reported in [222]. The value of C_n^2 along the propagation path can be calculated by applying the temperature structure function D_{Te} (L) and is given by [100]:

$$C_n^2 = \left(\frac{dn_1}{dT}\right) C_T^2, \qquad (5.1)$$

$$D_{T_e}(L) = \left\langle (T_a - T_b)^2 \right\rangle = \begin{cases} C_T^2 l_0^{-4/3} L^2, & 0 \leq L \ll l_0 \\ C_T^2 L^{2/3}, & l_0 \ll L \ll L_0 \end{cases}, \qquad (5.2)$$

where T_a and T_b are the temperature of two points separated by a distance L and C_T^2 is the temperature structure constant.

In order to anticipate the Rytov parameter σ_N^2 in the indoor atmospheric chamber, it is presumed that C_n^2 is constant between two measured temperature points. In our case, the values of the temperature are measured at four positions as shown in Figure 5-4. The scintillation index is predicted using (3.3) in which the values of C_n^2 are calculated using (5.2) and (5.1). The measured values are obtained from the received signal utilizing the process discussed in (3.25). It is important to mention that due to the high sensitivity of σ_N^2 to the temperature gradient, exact matching is difficult to obtain. For instance, a ±1°C change in temperate T_1 can causes ~ 2.5 time difference in the predicted scintillation index values. Note that for outdoor terrestrial FSO systems, the refractive index structure parameter C_n^2 is pretended to be constant and (3.3) is usually applied to approximate the log irradiance variance. However, in real systems, the temperate gradient $\Delta T / \Delta L$ can vary along the propagation length as the atmospheric conditions greatly vary over time and space. Consequently, the atmospheric link segmentation approach can be employed as an alternative method to characterise the FSO link.

5.5. The Controlled Fog Simulation Chamber

In [49,160], the measurements indicate that the occurrence of fog begins when the relative humidity (H) of the ROA is nearly 80 %. The density of the resulting fog reaches 0.5 mg/cm^3 for $H > 95\%$. Therefore, for high water vapour concentration conditions, the water condenses into tiny water droplets of radius 1– 20 μm in the atmosphere. It is possible to simulate fog in the lab by achieving H values that are close to 95%. Thus, artificially generated steam fog can be produced to mimic the ROF as presented in Figure 5-6.

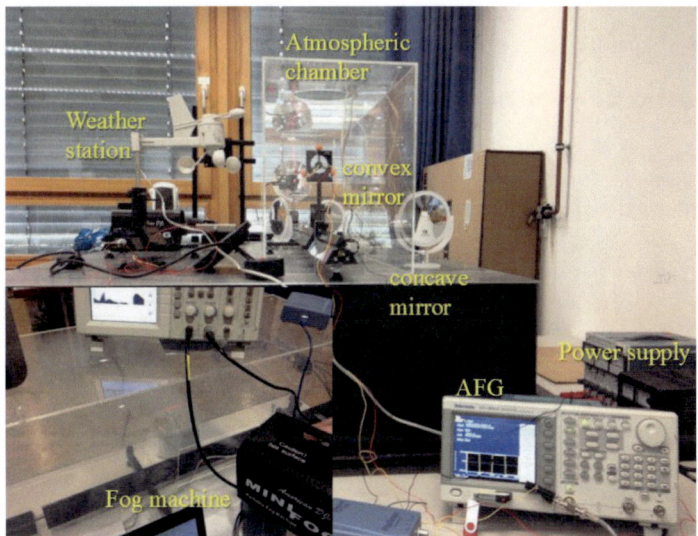

Figure 5-6: The laboratory controlled fog and FSO link setup.

From the time series analysis of the ROF, it is obvious that due to the rigid and very robust variation in outdoor FSO atmospheric circumstances, the fog is rather dynamic in size and heterogeneous along the length of the FSO link [161]. Usually, owing to this to fact, the characterisation of fog is based on the measured visibility V (km).

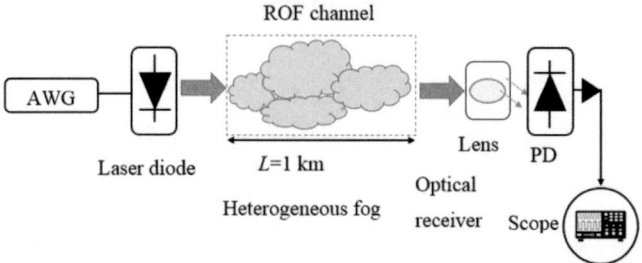

Figure 5-7: ROF scenario considering an FSO link length of 1 km.

Additionally, to contrast the physical resemblance of the lab generated fog with the ROF, the ROF attenuation data from field experiments at Prague published in [217] and the Czech Metrological Institute [179] are compared with the obtained laboratory-based fog attenuation data. The mean attenuation data from Prague for $0 < V < 1$ km and Nadeem et al. for $V < 0.6$ km depict a very good agreement with the measured lab data, as illustrated in Figure 5-8. This confirms that the physical characteristics of lab generated fog correspond to those of the ROF.

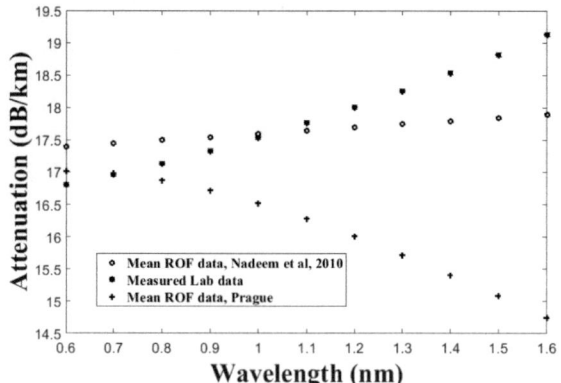

Figure 5-8: Comparison between the measured mean ROF attenuation and the mean lab generated fog attenuation.

However, to simulate steam fog, the humidity should be approximately 100%. In Chapter 7, steam fog is used in our lab based indoor atmosphere chamber to evaluate the

FSO link performance in a controlled fog chamber. In the case of a real FSO link, which is usually longer than 1 km, the spatial heterogeneity of the fog densities can vary from one position point to some 100 metres aside, as presented in Figure 5-7. As an outcome of this, the measured fog induced attenuation along the FSO link as well as the related visibility V data obtained from a visibility device (transmissometer) at a defined position are subject to significant fluctuations. This effect can be observed in the measured data at 830 nm for two fog levels as in [162]. Various transmissometers can be applied sequentially in steps of some hundred meters to minimise this error in measuring V, as demonstrated in [218]. Nevertheless, this method is complex and expensive to implement for a long range FSO systems. Therefore, the designed atmospheric chamber represents a simple approach which is applied to control and produce homogeneous fog conditions to measure V by employing several laser wavelengths within the length of the chamber, as shown in Figure 5-9. Hence, it allows measuring the values of the fog-induced attenuation corresponding to the actual V over the length of the FSO Link.

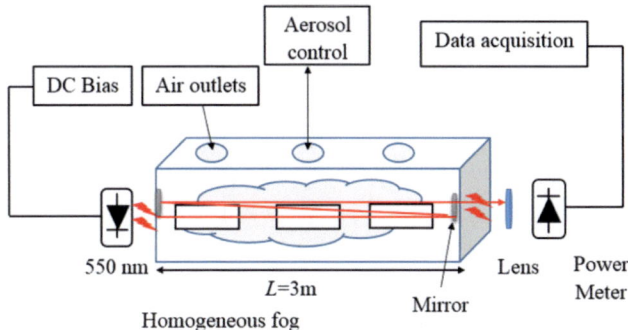

Figure 5-9: Block diagram of the experimental setup used to measure V over the length of the FSO link.

Achieving the homogeneity depends on various factors. First of all, the most important requirement for spreading fog is to have a proper enclosure of the chamber. On the other hand, in order to produce different fog densities, vents are used along the chamber. Proper spreading of fog facilitates measuring the exact values of V and obtaining good correlation of appropriating measured fog attenuation for the transmitted wavelengths of an FSO system. The fog is circulated homogenously by applying a combination of fans along the chamber.

Two different approaches were applied in the experimental work to characterise the fog induced attenuation. In the first approach, an optical power meter and a number of individual laser sources at wavelengths of 0.55, 0.67, 0.83, 1.31 and 1.55 µm were used with average transmitted optical powers P_T of -3.0 dBm, 0 dBm, 10 dBm, 6.0 dBm and 6.5 dBm respectively. In the second approach, a Rohde & Schwarz Q8341 Optical Spectrum Analyzer (OSA) with a spectral response ranging between 0.6 to 1.75 µm was used to obtain the attenuation profile. Following the first approach, the received optical power P_R without fog is measured for the reference as well as for normalization at the optical receiver R_x by using the power meter at each transmission wavelength (0.55, 0.67, 0.83, 1.31 and 1.55 µm). After that, a defined amount of fog is injected into the chamber permitting it to homogenously spread along the chamber. A time duration of 30 seconds is allowed for fog particles to settle homogeneously within the chamber before initiating the data acquisition (DAQ) process. The second step is to measure the received optical power P_R in the presence of fog, at a time step of 1 second for every wavelength. The normalized transmittance or loss is then calculated from the average received power with fog to the average received power with no fog. The fog induced attenuation β_λ (in dB/km) is measured using (5.3) and represents the measured loss at each transmission wavelength from light to dense fog conditions.

$$\beta_\lambda = -\frac{Loss}{4.343L}. \tag{5.3}$$

The link V is measured along the length of the chamber by applying (7.1) at a wavelength of 0.55 µm and comparing the measured value with the visibility threshold. The acquisition of the data is concluded when the visibility is equal to the threshold value (see Figure 5-10).

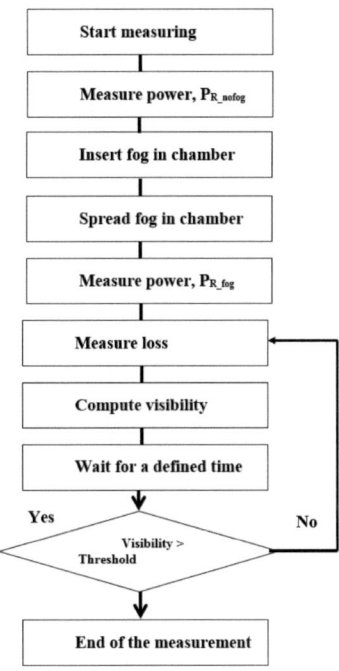

Figure 5-10: A block diagram for the measurement of fog induced attenuation and link visibility.

5.6. Summary

This chapter presented the design of an indoor atmospheric chamber and discussed approaches to control and simulate the atmospheric environment, such as turbulence and fog, as requirements to simulate the ROA as realistically as possible. The atmospheric chamber was used to mimic the outdoor FSO system characterisation process, and to perform measurements in homogeneous turbulence and fog conditions. Various methods to produce and control turbulence inside the atmospheric chamber were also outlined. Further investigations of the performance of FSO systems built on the atmospheric chamber under turbulence conditions and fog conditions will be conducted in Chapter 6 and Chapter 7 respectively.

6. PERFORMANCE OF FSO LINKS UNDER CONTROLLED ATMOSPHERIC TURBULENCE

6.1. Introduction

A number of methods can be used to combat the effect of turbulence, including utilising schemes such as multiple input multiple output (MIMO) [41, 119], temporal and spatial diversity [33] and aperture averaging [157,165]. However, selecting a modulation format that is immune to the scintillation effect is also of huge importance [43, 166]. In the turbulent atmosphere, data recovery employing a fixed threshold level is not an optimal choice when using the OOK scheme. Although adaptive threshold detectors can significantly improve the performance in turbulence [167], utilising them is not practicable because they demand adaptive optical components as well as permanent monitoring of and adaptation to the atmospheric conditions [168]. Alternatively, modulation techniques such as SIM and PPM, which are immune to turbulence induced amplitude fluctuations, could be applied [44, 143]. The BPSK-SIM scheme, which does not require an adaptive threshold, profits from a mature RF technology and requires a simple and low-cost direct detection receiver. However, BPSK-SIM requires a higher average transmission power than OOK, due to the DC bias requirement and the likelihood of signal distortion and signal clipping [49, 145].

It has been reported in numerous research studies that in cases of high background noise, M-PPM is considered to be the optimum modulation scheme in a Poisson-type channel [219], [220]. With the increase in the order of M in M-PPM, the robustness against background radiations increases even further due to its low duty cycle and the shorter integration interval of the photodiode. For PPM, the increase in atmospheric scintillation leads in an increase in the required signal power level to attain a given BER. Increasing the signal strength can be used to minimise the scintillation effect at a low scintillation index, but as turbulence strength increases, the BER performance tends towards high BER asymptotic values [1]. SIM schemes modulate the frequency and phase of the RF carrier and do not require dynamic thresholding for optimal detection. Phase fluctuations are less marked in turbulent atmosphere, and therefore SIM offers better performance in comparison to OOK [221].

In this chapter, an experimental characterisation of the turbulence effect on the performance of FSO links employing different modulation techniques is reported. The experiment is carried out in a controlled laboratory environment, where turbulence is generated in a dedicated indoor atmospheric chamber.

6.2. BPSK-SIM under Controlled Turbulence

The experimental LOS FSO link used for data transmission through a turbulence channel is shown in Figure 6-1 and Figure 6-2. A narrow divergence laser beam at a wavelength of 830 nm was used as the FSO transmitter. The emitted beam intensity was modulated by a data source generating a stream of $2^{13}-1$ pseudo random bit sequence, which was either Ethernet (10BASE-T) or Fast-Ethernet (100BASE-SX). The laser was appropriately biased and modulated.

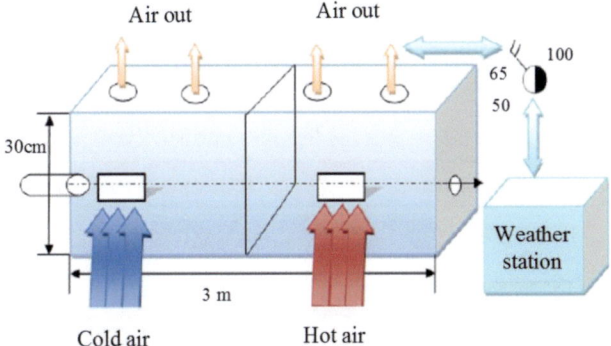

Figure 6-1: The laboratory turbulence chamber which is excited by hot and cool air.

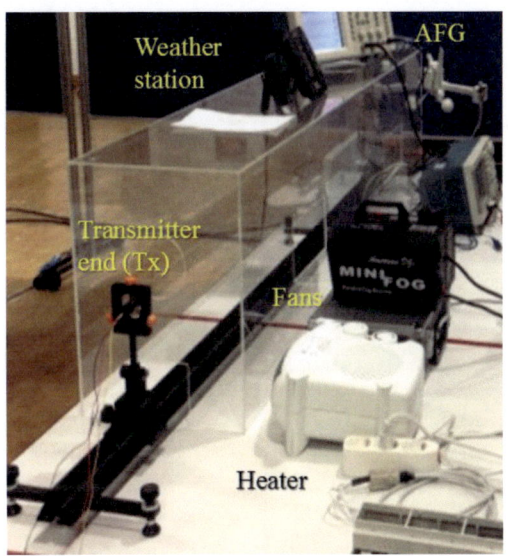

Figure 6-2: The chamber and FSO link setup in the laboratory.

The laboratory atmospheric channel is a closed glass chamber with 300×30×30 cm³ dimensions, as described in Figure 6-1 and Figure 6-2. The temperature and wind velocity conditions in the chamber were controlled to simulate atmospheric conditions as closely as possible; adopting the control approach described in Chapter 5. For a strong turbulence regime, multiple scattering effects must be considered, which are not included in (3.15). Therefore, an improved model should be used [47]. Table 6-1 relates the turbulence strength with Rytov parameter [47]. In this experimental study, measurements for the weak and medium turbulence conditions were carried out.

Table 6-1: Turbulence strengths and the corresponding Rytov parameter.

Turbulence	Rytov parameter
Weak	$\sigma_I^2 < 0.3$
Medium	$\sigma_I^2 \approx 1$
Strong	$\sigma_I^2 \gg 1$

As the optical beam propagated through the chamber, it experienced different atmospheric turbulence levels before being collected at the receiver. The receiver front-end consisted of an optical concentration lens and a PIN photodetector. The equivalent photocurrent at the output of the photodetector was amplified using a trans-impedance amplifier integrated circuit (IC), and the recovered data was used to determine the bit error rate (BER) performance. All relevant parameters for the designed chamber are given in Table 6-2.

Table 6-2: Main parameters of the turbulence chamber.

Parameter	Value
Dimension	300×30×30 cm^3
Temperature range	20 - 80 C$^\circ$
Wind speed	4 - 5 m/s

The level of turbulence strength was controlled by localising the same heating source near and far away from the FSO transmitter. The ray tracing diagram in Figure 6-3 and Figure 6-4 illustrates this concept. The optical beams shown in both Figure 6-3 and Figure 6-4 approximately experience the same degree of deflection due to the same level-controlled turbulence source being used. However, due to the geometrical configuration, lower power levels will be collected at the receiver as shown in Figure 6-3.

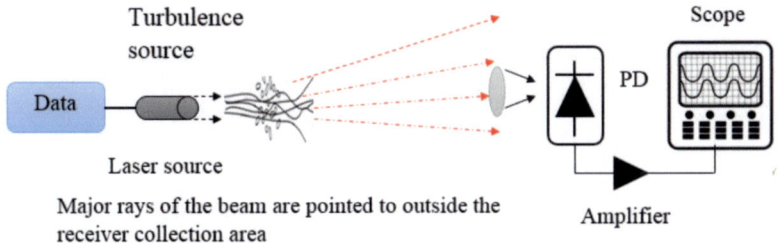

Figure 6-3: Sketch of diverted beams due to the turbulence source being positioned near the transmitter.

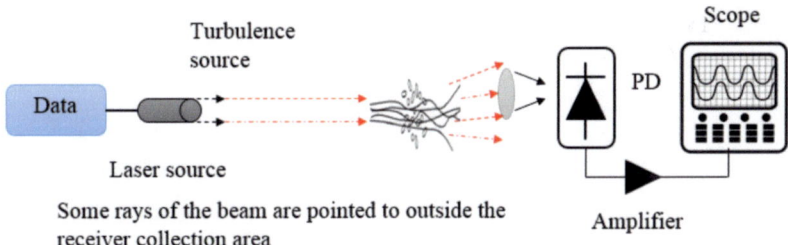

Figure 6-4: Sketch of diverted beams due to the turbulence source being positioned near the receiver.

The purpose of this demonstration was to investigate the BER performance of an FSO link under the influence of atmospheric turbulence. To ensure valid comparisons, measurements carried out in the laboratory environment were taken in similar environmental conditions as much as possible. Data packets with a length of ~ 600 - 1000 bytes, typically used in the Gigabit Ethernet (GbE) data networks, were used in the experiment. Further details of the setup and parameters used for the demonstration are shown in Table 6-3.

Table 6-3: Parameters of the FSO link demonstration.

Parameter		Value
Data source	PRBS length	$2^{13} - 1$
	Format	NRZ / RZ
	Modulation voltage	LVDS (400 mV$_{pp}$)
Laser diode	Peak wavelength	830 nm
	Maximum optical power	10mW
	Class	IIIb
	Beam size at aperture	5mm × 2 mm
	Beam divergence	5 mrad
	Modulation bandwidth	75 MHz
Photodetector	Wavelength at maximum sensitivity	900 nm
	Spectral range of sensitivity	750 - 1100 nm
	Active area	1 mm2
	Half angle field of view	± 75 Deg
	Spectral sensitivity	0.59 A/W
	Rise and fall time	5 ns
	Reversed bias voltage	40 V
Lens	Diameter	25 mm
	Focal length	200 mm
Receiver	Transamplifier (IC)	LT6202
	Bandwidth	240 MHz
	Transimpedance amplifier gain	15 kΩ

Figure 6-5 depicts the block diagram of a BPSK-SIM system with two subcarriers. Appling BPSK, the input data $\{j_1, j_2\}$ is modulated onto the RF subcarriers, whose amplitudes, frequencies and phases are $\{a_{c1}, a_{c2}\}$, $\{\omega_{c1}, \omega_{c2}\}$ and $\{\varphi_{c1}, \varphi_{c2}\}$ respectively. In Figure 6-6, $g(t)$ represents the pulse shaping function. The combination of the two RF subcarrier signals is then employed to modulate the intensity of the optical source. Before that, however, a D.C. signal b_0 is added to the composite RF signal, to ensure that the optical source is suitably biased at the centre of its linear dynamic range, so as to fit the full swing of the sinusoidal subcarrier signal.

In the atmospheric channel, turbulence effects are modelled utilizing the log-normal induces the intensity of the transmitted optical signal to fade. At the receiver, the composite SIM signal, superimposed on the envelope of the incoming optical signal, is recovered via the direct detection (DD) scheme. Electrical bandpass filters are employed to filter the individual subcarriers $i_{bi}(t)$, followed by a standard RF coherent detector to recover the transmitted data sequence $\{\hat{j}_1, \hat{j}_2\}$. The system noise (thermal and shot) is modelled as an AWGN and no inter-symbol interference is accounted for since the link under consideration is a direct line of sight with no multipath propagation, as described in Figure 6-5 [1].

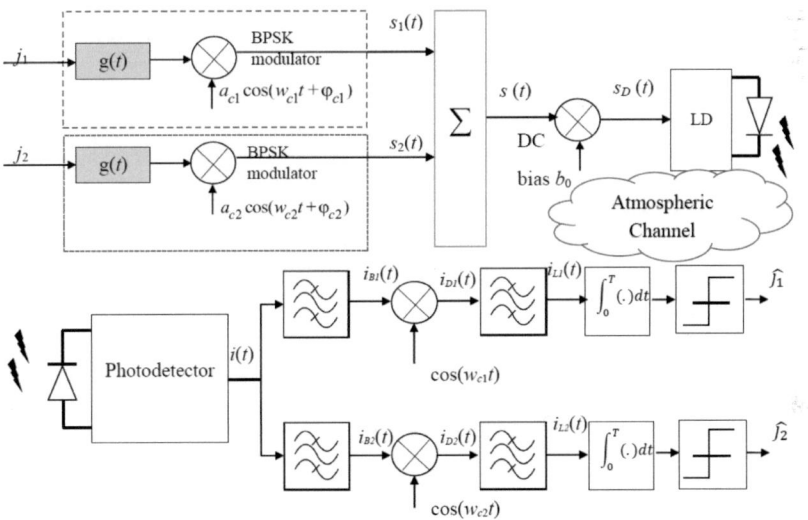

Figure 6-5: The block diagram of a subcarrier BPSK system with two subcarriers.

In the experiment, turbulence was generated inside the chamber by pumping hot air either through the vents near the transmitter, in the middle of the chamber or near the receiver. Table 6-4 shows the measured values of σ_l^2 at these positions. The shot noise variance is excluded from the reported σ_l^2 values in the table. Notice that by using the same turbulence source and varying its position along the chamber, we could generate different levels of turbulence (i.e. σ_l^2) according to the concept outlined in Figure 6-4 and

Figure 6-5. The concept is also valid for actual outdoor FSO systems. For the measured σ_I^2 value of 0.1, the turbulence generated in the chamber could be equivalently considered as a weak turbulence for the outdoor environment. A histogram of the received signal affected by medium turbulence was obtained and is plotted in Figure 6-6. The distribution of the received signal has a Gaussian profile when there is no turbulence (see Figure 6-6 (a)) as the Gaussian-distributed shot noise (due to the ambient noise) is predominant. However, the received signal's pdf profile changes to a log-normal profile when turbulence is introduced at the centre, and at the front end of the chamber (see Figures 6-6 (b) and (c)).

Figure 6-6 (a-c) shows the histogram of the received signal for a bit '1' BPSK sequence with and without turbulence. Notice that the total number of occurrences is normalized to unity, which represents the occurrence of pdf. Also illustrated in Figures 6-6 are the curve-fitting plots using the Gaussian and lognormal models. With no turbulence, the distribution is Gaussian, whereas with turbulence, the pdf is best fitted with a log-normal distribution, as shown in Figures. 6-6 (b, c).

Table 6-4: Measurement results of the turbulence strength inside the chamber.

Turbulence position	Near transmitter	Middle of the chamber	Near receiver
Rytov variance	$\sigma_I^2 = 0.09$	$\sigma_I^2 = 0.12$	$\sigma_I^2 = 0.08$
Turbulence strength	Weak	Weak	Weak

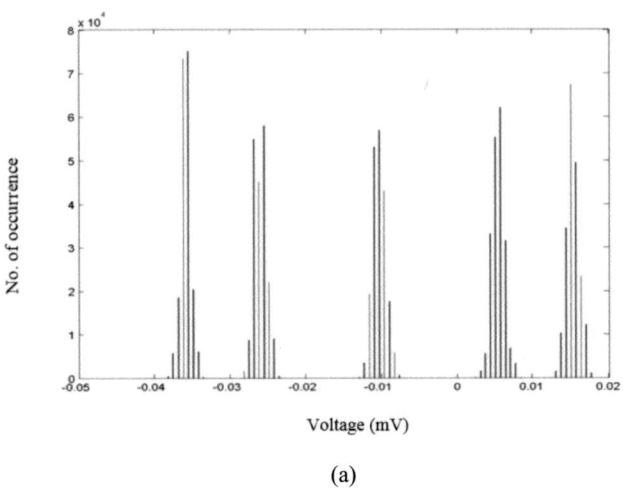

(a)

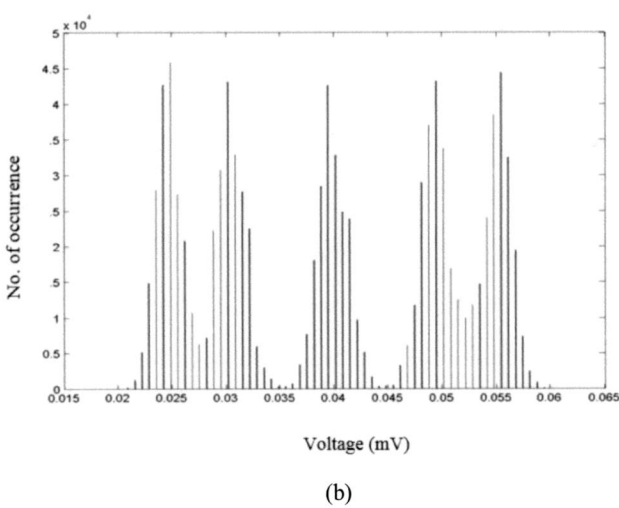

(b)

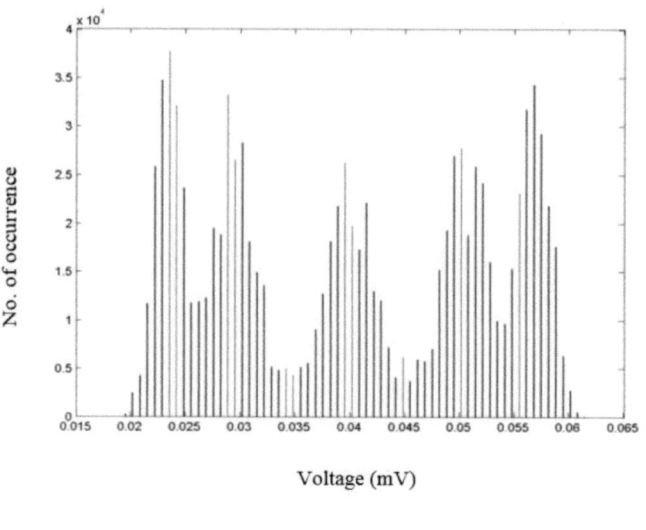

(c)

Figure 6-6: Histogram of the received signal in cases of: (a) no turbulence, (b) turbulence in the middle of the chamber, and (c) turbulence near the transmitter, for weak turbulence ($\sigma_l^2 = 0.12$).

The eye diagrams of the received signal for the BPSK data format at 50 Mbit/s data rate are depicted in Figure 6-7. The modulation input voltage was LVDS (400 mVpp). It is noticed that, the top (bit '1') and base (bit '0') levels of the received signals vary at a much wider margin when turbulence is introduced. This results in the reduction of the measured Q-factor and hence the BER performance.

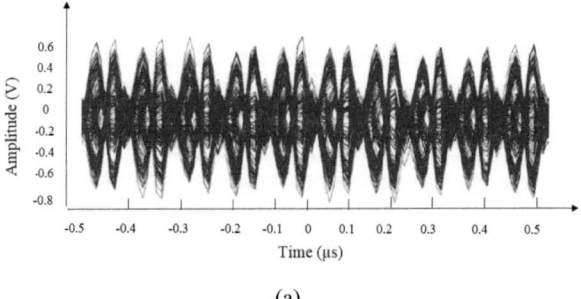

(a)

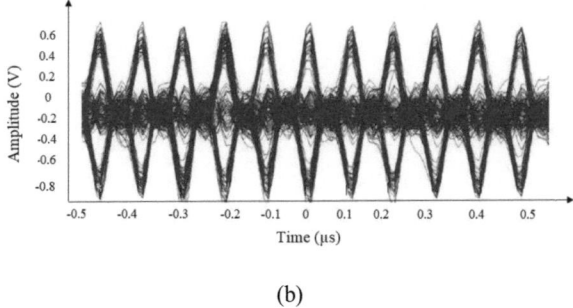

(b)

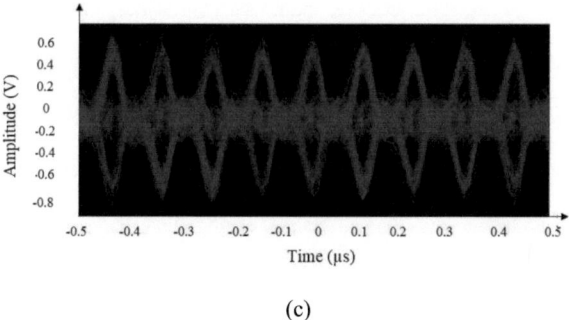

(c)

Figure 6-7: The measured eye diagram of the received BPSK signal in cases of (a) no turbulence and a data rate of 50 Mbit/s, (b) weak, middle of the chamber turbulence with $\sigma_l^2 = 0.1$ and a data rate of 50 Mbit/s, and (c) near receiver turbulence and near transmitter turbulence with $\sigma_l^2 = 0.09$.

The eye-diagrams clearly show the superior performance results of BPSK to those reported in [12], as the height of the eye-opening is nearly identical with and without turbulence. This is because the information in BPSK schemes is encoded in the phase of the subcarrier signal rather than the amplitude, as in the OOK data format. This makes BPSK signals robust against turbulence, since the turbulence effect induces severe fluctuations in the intensity of the signal (e.g. NRZ signals), rather than the phase. The advantage of the BPSK scheme also include the cost of power and bandwidth efficiencies.

6.3. Hybrid BPSK-SIM- PPM FSO with Turbulence

6.3.1. Introduction

A new hybrid PPM-BPSK modulation scheme is presented and its performance is simulated. The results are then compared with the PPM and BPSK-SIM modulation techniques. In addition, experimental results for the 4-PPM and BPSK-SIM techniques are also reported [169]. Furthermore, numerical simulation results for the BER performance of PPM-BPSK in a lognormal atmospheric channel for a range of turbulence variance are presented. The results are compared with those of the BPSK-SIM and 4-PPM schemes, and show that the performance of the hybrid PPM-BPSK scheme is superior to that of BPSK while having the same bandwidth requirement. However, for all levels of turbulence, the performance of the proposed hybrid PPM-BPSK scheme is inferior to that of PPM.

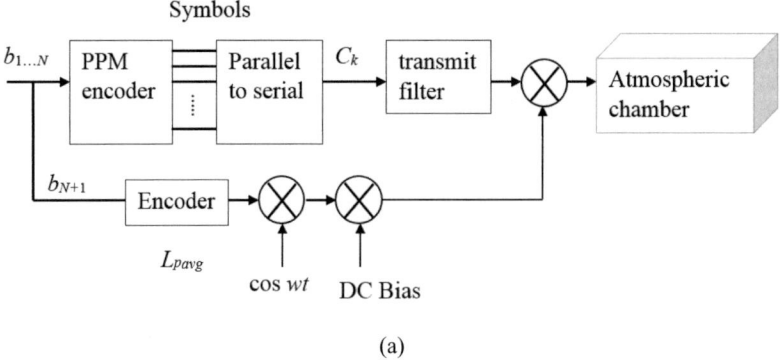

(a)

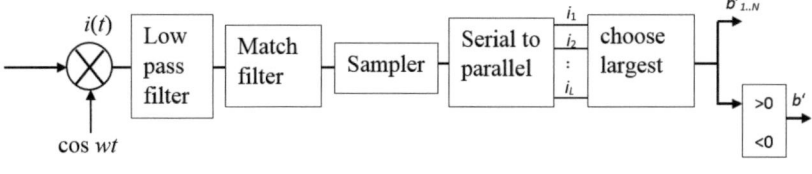

(b)

Figure 6-8: Block diagram of an L-PPM-BPSK-SIM FSO system showing the building blocks of the (a) transmitter, and (b) receiver [169].

At the receiver, choosing the index of the slot containing the maximum of the signal in a symbol, the conditional slot error rate (SER) is obtained by [1, 39]:

$$P_{eslot-con} = Q\left(\frac{I}{\sqrt{2\sigma_{ppm}^2}}\right), \qquad (6.1)$$

where the noise variance $\sigma^2 = N_0 R_b q / (2\log_2 q)$, I is the intensity of the received signal with a mean value of $I_o = qP_{av}$ [170], N_0 is the double side spectral density of the Gaussian noise, and P_{av} is the average transmitted optical power. By averaging the conditional slot error rate over turbulence, the unconditional slot error rate is obtained by:

$$P_{eslot-erro} = \int_0^\infty Q\left(\frac{I}{2\sigma^2}\right) P_I(I) dI. \qquad (6.2)$$

The main disadvantage of the PPM technique is its low bandwidth efficiency. To solve this, PPM-BPSK is proposed. PPM-BPSK is a combination of PPM and BPSK, in which a block of $N+1$ bits of input information is selected. The first N bits are modulated by the position of the pulse, and the last bit is modulated via the intensity of the pulse. The block diagram of this proposed scheme is illustrated in Figure 6-8. In PPM-BPSK, each symbol consists of an optical pulse in one slot duration T_s within 2^N slots, and its intensity is modulated via BPSK-SIM. As illustrated in Figure 6-10 (a), the transmitter consists of a PPM encoder, a parallel to serial convertor, a transmitter filter with an impulse response of $g(t)$, and a BPSK encoder. The PPM encoder codes the first N-bits of input data, and after converting into a serial form (consisting of only one '1' and q-1 zeros), the bit stream is passed through the transmitter filter to ensure the PPM signal. The $(N+1)^{th}$ data bit is encoded by the BPSK encoder into '1' or '-1', and is modulated by an RF signal. A DC bias is then added to ensure a positive BPSK signal. The intensity of this signal is finally modulated by the PPM signal to obtain a PPM-BPSK-SIM modulated signal. Thus, the current signal in the receiver can be expressed by:

$$i(t) = RI\left[\sum_{k=0}^{L-1} a_c c_k (1+\cos(wt))g(t-kT_g))\right] + n(t), \qquad (6.3)$$

where R is the responsivity of the photodetector, c_k is the codeword of PPM, $L = 2^N$, $T_g=(N+1)T_b/L$, $I_0=LP_{av}$, and $a_c=\{1,-1\}$ and refers to the $(N+1)^{th}$ bit. At the receiver, the photocurrent is down converted via a low pass filter, a match filter, a sampler, and the reference carrier signal cos (wt). Demodulation is carried out by finding the largest absolute signal in one symbol period, where its position within the symbols determines the N-bit received data.

6.4. Results and Discussion

6.4.1. Simulation Results

The BPSK, SIM, and PPM modulation schemes described above were simulated in MATLAB. The simulations were carried out for $N = 1$ in hybrid PPM-BPSK, and $q=4$ in PPM, where $T_g = T_b = 4T_s$. To have a fair comparison between the different modulation schemes, the average transmitted power and R_b were fixed. All other simulation parameters are shown in Table 6-5. Figure 6-11 shows the predicted BER against SNR for 2-PPM with and without turbulence for a range of Rytov variance values $\sigma_l^2 = \{0.1, 0.3, 0.5\}$, where $P_{slot\_error} = P_b$. In addition, Figure 6-12 depicts the BER vs. SNR curve predicted for BPSK with and without turbulence for a range of Rytov variance values. A good agreement is obtained between the simulations and the theoretical results, thus validating the simulation results for both BPSK and 2-PPM. As shown in Figures 6-11 and 6-12, increasing the turbulence level decreases the system performance as expected. Figures 6-13, 6-14 and 6-15 plot the simulated BER results versus SNR for PPM-BPSK, BPSK and 4-PPM, for a range of Rytov variance values of $\sigma_l^2 = 0.1, 0.3$ and 0.5 respectively. As illustrated in the figures, the performance of PPM-BPSK is superior to that of BPSK while having the same bandwidth, but its performance is inferior to that of 4-PPM for all simulated turbulence levels.

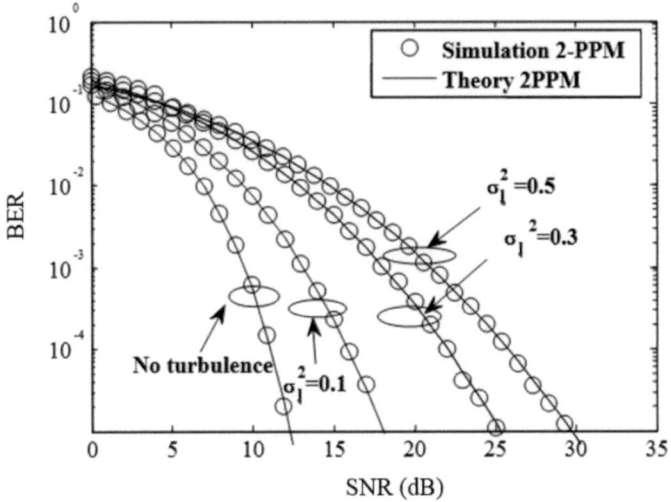

Figure 6-9: Simulated and predicted BER against SNR for 2-PPM with and without turbulence for a range of Rytov variance values.

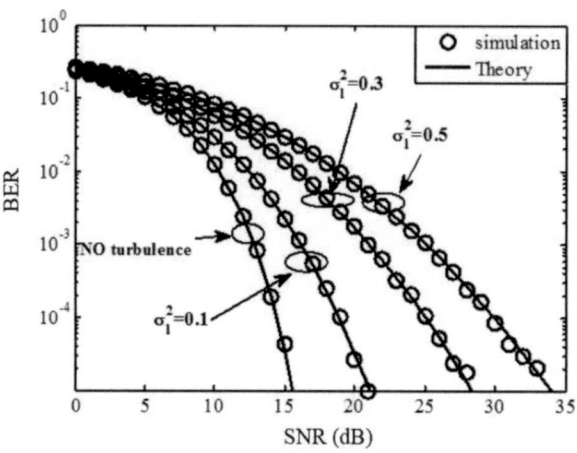

Figure 6-10: Simulated and predicted BER against SNR for BPSK-SIM with and without turbulence for a range of Rytov variance values.

Table 6-5: The simulation parameters.

Parameter		value
Data rate R_b		1 Mbps
Carrier frequency		4 MHZ
Sampling frequency		20 MHZ
Laser wavelength λ		850 nm
Responsivity R		1 A/W
Modulation index ξ		1
Butterworth Low pass filter	Order	6
	Bandwidth	1 MHZ
N in PPM_BPSK		1
q in PPM		2, 4
Pulse shaping		Rectangular

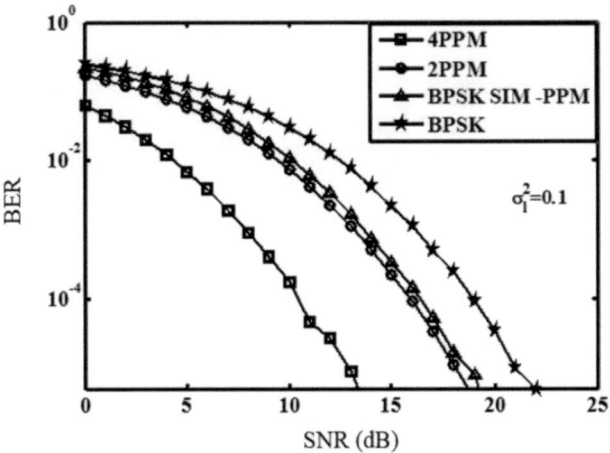

Figure 6-11: Simulated BER versus SNR for 4-PPM, BPSK and PPM-BPSK with turbulence for a Rytov variance of $\sigma_I^2 = 0.1$.

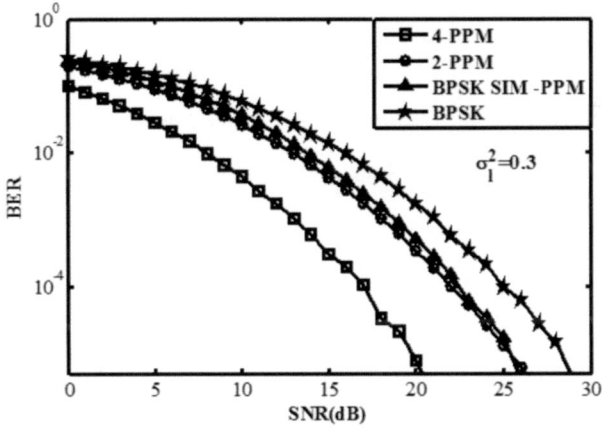

Figure 6-12: Simulated BER versus normalised SNR for 4-PPM, BPSK and PPM-BPSK with turbulence for a Rytov variance of $\sigma_I^2 = 0.3$.

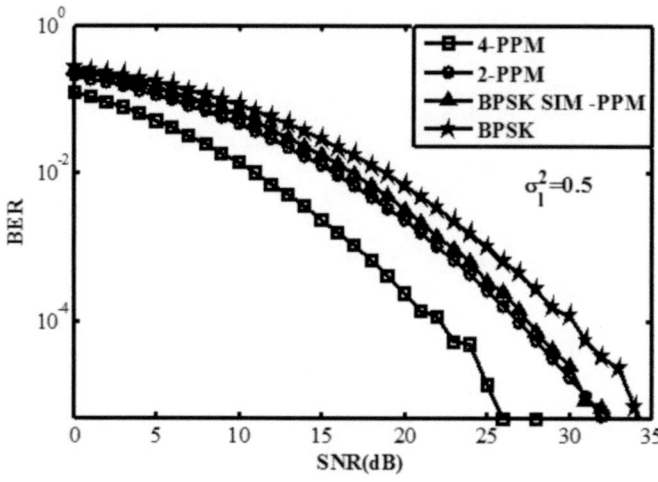

Figure 6-13: Simulated BER of 4-PPM, BPSK and PPM-BPSK against normalised SNR with turbulence for a Rytov variance of $\sigma_l^2 = 0.5$.

6.4.2. Experimental Results

A typical FSO link consists of a transmitter and a receiver separated by the atmospheric channel. The experimental set-up for the controlled study of the scintillation effect on the FSO link performance for the BPSK-SIM and 4-PPM modulation schemes utilised a laser source with a maximum optical output power of 10 mW at a wavelength of 830 nm. The intensity of laser was varied according to the modulating data format. To ensure the linearity of the system, the laser was accurately biased and the peak-to-peak voltage of the input signal was kept within the specified values. The receiver front-end consisted of an optical telescope (or lens) and a photo-detector. The electrical signal at the output of the PIN-photo-detector was amplified using a trans-impedance amplifier.

The experimental data for the two different modulation schemes were recorded under a controlled weak turbulence environment and analysed using eye diagrams and the received signal distributions. The eye-diagrams provide a good indication of the quality of the received optical signal at the end of the turbulence channel. The measured eye-diagrams for the received signals are depicted in Figures 6-16, 6-17 and 6-18. The height of eye-opening is smaller in the presence of turbulence due to the turbulence-induced intensity fluctuations. The height of eye-opening is greater for the 4-PPM signal compared to the BPSK-SIM signal in the presence of turbulence, indicating that 4-PPM is less sensitive to intensity fluctuations under weak turbulence conditions. In Table 6-6, it is indicated that the 4-PPM signalling format offers a superior Q-factor performance compared to BPSK-SIM for $\sigma_I^2 = 0.1$, where approximately ~1.65 higher Q-factor values are recorded for 4-PPM compared to BPSK-SIM.

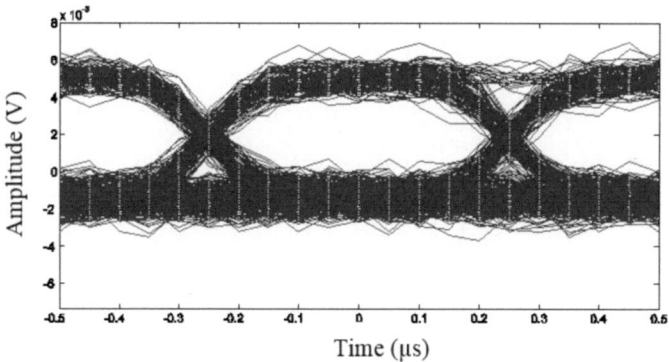

Figure 6-14: Eye diagram representations of 4-PPM for a 200 mV$_{p-p}$ input signal with turbulence $(\sigma_I^2 = 0.1)$.

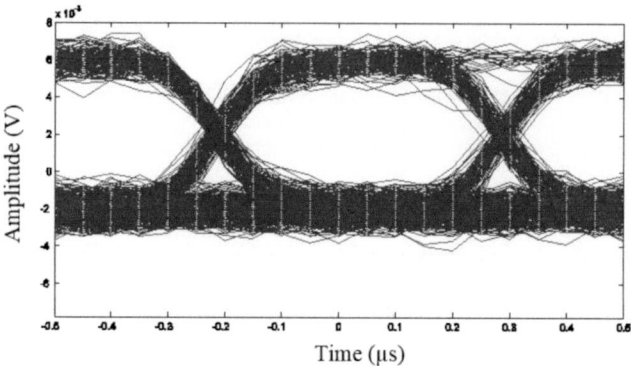

Figure 6-15: Eye diagram representations of 4-PPM for a 400 mV$_{p-p}$ input signal with turbulence $(\sigma_I^2 = 0.1)$.

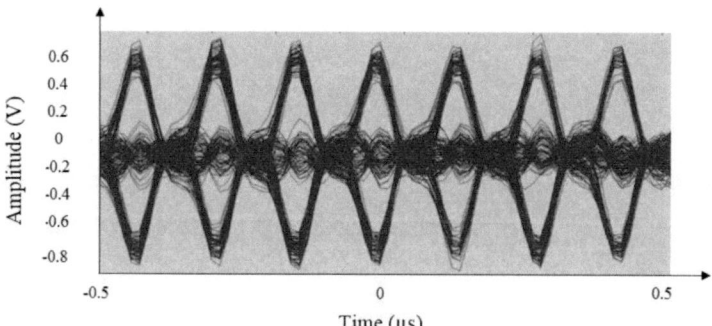

Figure 6-16: Eye diagram representations of the received BPSK signal with turbulence $(\sigma_I^2 = 0.1)$, a data rate of 50 Mbit/s, and a 400 mV$_{p-p}$ input signal.

Table 6-6: Experimentally measured Q-Factor values.

4-PPM Q-Factor	BPSK Q-Factor
2.41	1.43
2.52	2.09
5.97	3.6
6.91	4.04
7.14	4.25
7.23	5.2

6.5. Summary

This chapter presented an investigation of the performance of FSO links under atmospheric turbulence channels. A dedicated laboratory atmospheric chamber was used to investigate the effect of temperature induced turbulence on FSO links performance. Methods to generate and control turbulence were discussed and practically demonstrated. The obtained data demonstrated that weak turbulence can severely affect the link performance. Furthermore, the results showed that the turbulence effect is also dependent on the data format adopted to directly modulate the laser source, and that the profile of the received signal distribution after turbulence is Gaussian rather than log-normal. The results also demonstrated that the BPSK scheme is less sensitive to irradiance fluctuations under weak turbulence conditions. Moreover, a new modulation scheme, named PPM-BPSK, was proposed in this chapter. The performance of PPM-BPSK ($N = 1$), 4-PPM and BPSK-SIM was investigated and compared. The results showed that PPM-BPSK offer similar performance to 2-PPM, a superior performance to BPSK-SIM while having the same bandwidth, but an inferior performance to 4-PPM for all turbulence levels. The experimental investigation indicated that 4-PPM is more robust to turbulence impairments in the FSO link compared to BPSK-SIM.

7. PERFORMANCE OF FSO UNDER CONTROLLED FOG CONDITIONS

7.1. Introduction

The constituents of the atmosphere, particularly fog, severely hinder FSO systems performance by scattering and absorbing photon energy [173, 174]. This consequently leads to reduced received optical power hence reduced BER performance [171,172]. Dense fog can potentially increase the BER, therefore preventing a high data rate FSO link from meeting the 99.999% link availability [175, 176]. A number of research works investigating the effect of fog on FSO systems have been conducted. These mostly focus on channel modelling and measurements of the attenuation [160, 177] as well as the performance of classical modulation schemes in the presence of fog [49]. However, more work applying various as well as combined modulation techniques over fog affected FSO links is still required in order to optimize the performance of FSO systems under attenuating fog conditions.

In this chapter, an experimental study is reported to test and mitigate the effect of fog by employing power efficient modulation schemes with different P_{opt} values. The system performance is theoretically and experimentally evaluated for BPSK-SIM, 4-PPM, QPSK, Hybrid 4-PPM-BPSK formats for Ethernet line data rates under light to dense fog conditions.

7.2. Coherent Detection of BPSK FSO Links with Fog

7.2.1. The Characterization of Fog Attenuation Uisng a Laboratory Chamber

a) Fog and visibility

In practice, V (i.e. the meteorological visual range) is used to characterise the fog attenuation over a transmission link. Applying Koschmieder law, the meteorological V (km) can be calculated from the atmospheric attenuation coefficient β_λ and the visual threshold at 0.55 μm wavelength as follows [210]:

$$V = -\frac{10\log_{10}(T_{th})}{\beta_\lambda}, \qquad (7.1)$$

where β_λ is normally expressed in (dB/km) and is mathematically defined by knowing the transmittance T of the optical signal and the propagation distance L using the Beer-Lambert law [198]:

$$\beta_\lambda = \frac{10\log_{10} T_{th}}{V(km)} \left(\frac{\lambda}{\lambda_0}\right)^{-q}. \qquad (7.2)$$

In general, due to the complexity involved in the physical properties of fog, such as the particle size and the non-availability of particle dispersion, the fog induced attenuation of an optical signal can be predicted using empirical models [15, 178, 179]. These models use the visibility data in order to estimate the fog induced attenuation. The empirical relationship, which relates V with the fog attenuation, is defined by the Kruse model [179]:

$$V(km) = \frac{10\log_{10} T_{th}}{\beta_\lambda} \left(\frac{\lambda}{\lambda_0}\right)^{-q}, \qquad (7.3)$$

where T_{th} is the transmission threshold, usually taken as 2% of the original optical power, λ_0 is the maximum spectrum of the solar band, and q is the coefficient related to the particle size distribution in the atmosphere. The Kruse model estimates the haze attenuation from visible–NIR wavelengths. However, Isaac I. Kim altered the Kruse model by applying theoretical suppositions for the fog by defining the q values [179]. Therefore, in this work, the Kim model is adopted for theoretical analysis, which indicates that the atmospheric attenuation coefficient β_λ for $V < 0.5$ km, is wavelength independent. Kim modified the Kruse model using theoretical assumptions for the fog by defining q as follows [15]:

$$q = \begin{cases} 1.6 & \text{for } V > 50 \text{ km} \\ 1.3 & \text{for } 6 < V < 50 \text{ km} \\ 0.16V + 0.34 & \text{for } 1 < V < 6 \text{ km} \\ V - 0.5 & \text{for } 0.5 < V < 1 \text{ km} \\ 0 & \text{for } V < 0.5 \text{ km} \end{cases} \qquad (7.4)$$

b) FSO links performance with fog

Assuming $I(f)$ and $I(0)$ are the intensity of the received optical signals with and without fog respectively, the transmittance T is given by the Beer-Lambert law [198] as:

$$T = \frac{I(f)}{I(0)} = \exp(-\beta_\lambda L), \qquad (7.5)$$

However, we want to correlate the Q-factor of the received signal for a given T for different fog conditions. The Q-factor, which represents the SNR at the receiver with no fog, is given by:

$$Q = T_0 \frac{|I_1 - I_0|}{\sigma_1 + \sigma_0}, \qquad (7.6)$$

where T_0 denotes the maximum transmittance, I_1 and I_0 are the average detected signal currents for bit '1' and '0' respectively, whereas σ_1 and σ_0 are the standard deviation of the noise values for bit '1' and '0' respectively. With fog, and assuming that the ambient noise level does not change with fog density, the Q-factor can be estimated as follows:

$$Q_{fog} = T_{fog} Q, \qquad (7.7)$$

where T_{fog} is the transmittance measured in the presence of fog.

7.2.2. Experimental Setup

The setup used in the laboratory-based FSO experiment consists of an optical T_x and R_x separated by an atmospheric chamber, as shown in Figure 7-1(a). The controlled channel, represented by the atmospheric chamber, had dimensions of $300 \times 30 \times 30$ cm^3.

A typical FSO link setup is shown in Figure 7-1(b). A $2^{13}-1$ bit long pseudorandom data sequence (PRBS) of one-bit length BPSK modulation was used to intensity modulate the laser source at a data rate of 25 Mbps. The intensity modulated optical beam propagates through the atmospheric chamber, is reflected at a convex and a concave mirror, and is detected at the receiver which is composed of an optical concentration lens and a PIN photodetector integrated with a wide-bandwidth transimpedance amplifier. The fog was generated using the fog machine (water steam) with 100% humidity to replicate ROF in this study. The chamber had two compartments, each with a vent to allow air circulation using a fan. The fog was injected into the chamber employing a fog generator at a rate of 0.74 m³/sec.

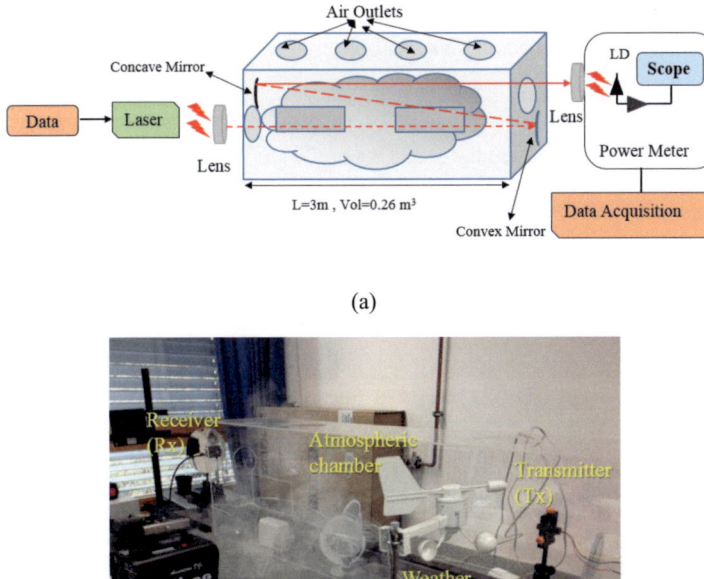

(a)

(b)

Figure 7-1: (a) Block diagram of the experiment setup, and (b) the laboratory controlled atmospheric chamber and FSO link setup.

The fog intensity in the chamber was controlled by a number of fans and ventilation inlet/outlets allowing us to manage the fog flux among the chamber homogeneously and control the visual contrast (transmittance) of the FSO link. The average received optical power; the Q-factor and BER of the received signal were simultaneously measured for different fog conditions (i.e. from low to high visibilities).

In order to measure the fog outcome, the average received P_{opt} was measured at both sides (R_x and T_x) before and after the introduction of the fog into the atmospheric chamber. The major contribution to the dispersion of the beam is the particle size [124], which is approximately close to the optical beam wavelength. In general, both fog and haze are major contributors to the Mie scattering due to the optical wavelengths being in the order of 1 - 15 μm. Due to the random nature and occurrence of fog as well as its type, its effect on FSO links performance can be readily measured and characterised by the visibility data [180, 181]. We evaluated the scattering coefficient β_λ using (7.5) corresponding to the measured T at 650 nm. The visibility V is measured using (7.4), utilising the wavelength independent model known as the Kim model for $V < 500$ m and for a range of transmittance values, as shown by Table 7-1 below.

Table 7-1: Measured T and visibility values.

Fog	Dense	Thick	Moderate	Light
T	< 0.31	0.31-0.71	0.71-0.8	0.8-0.9
V (m)	< 70	70-250	250-500	500-1000

Figure 7-2 illustrates the dependence of the Q-factor on the link transmittance for a range of optical power P_{opt} levels. Also depicted are the predicted results for P_{opt} =1.1 dBm, showing a close agreement with the measured data. Note that, in the region of $T < 14$ % (dense fog), the Q- factor values are nearly the same for all values of P_{opt}. This is due to the background noise being the dominant noise source (see the eye diagrams in Figure 7-3). To achieve a BER of 10^{-6}, the Q- factor value must be ~ 4.25. However, this Q-factor value cannot be achieved under the dense fog condition (i.e. T < 31%) even for the highest input power used (i.e. P_{opt} = 1.1 dBm) (see Figure 7-2). With dense fog, the link range drops to < 70 m with reduced link availability, thus making the FSO link less attractive. In such cases, one could increase the optical power provided it is kept below the eye safety level or switch to a lower data rate RF technology to maintain the maximum availability of 99.999%. For the thick fog condition ($31 < T < 70\%$), increasing P_{opt} from 0.7 dBm to 1.1 dBm results in increasing the Q-factor from 2 to 7.6, thus corresponding to a BER improvement from 10^{-3} to $>10^{-9}$.

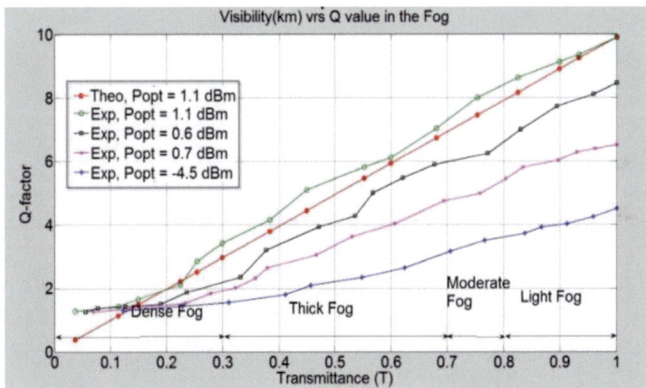

Figure 7-2: A plot of the Q-factor against the transmittance T.

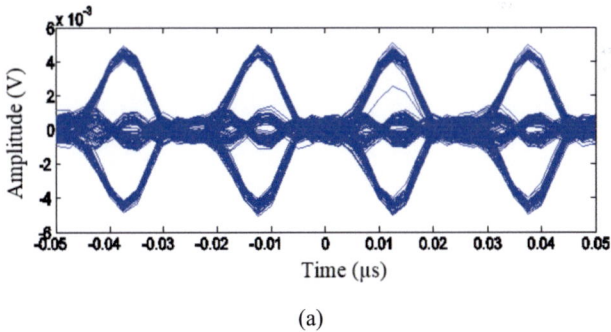

(a)

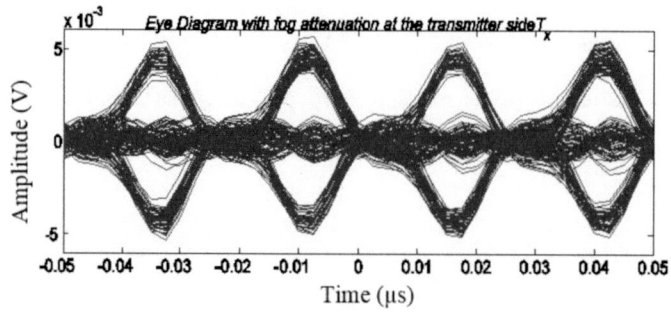

(b)

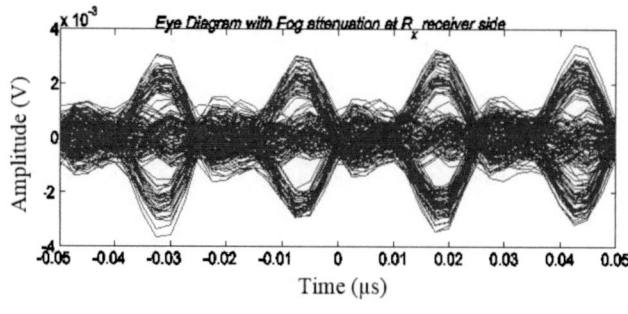

(c)

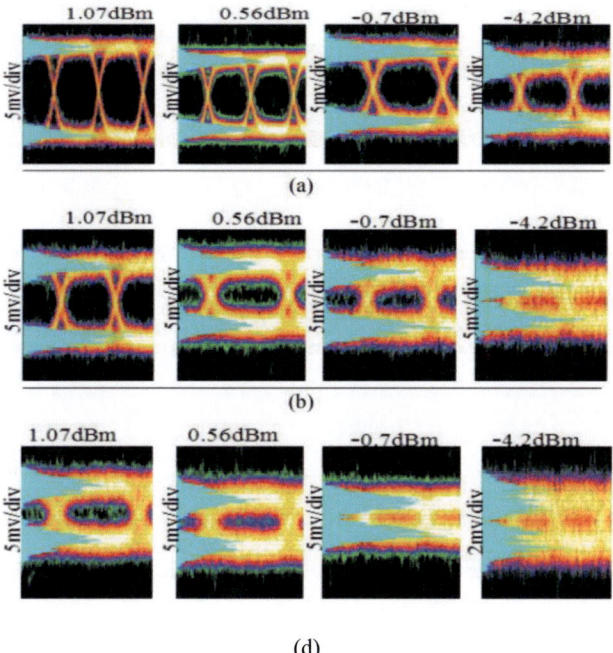

(d)

Figure 7-3: The measured eye diagram of the received BPSK signal with (a) no fog conditions, and a data rate of 25 Mb/s, (b) fog attenuation at the transmitter side, and (c) fog attenuation at the receiver side (x-axis in ns), and (d) OOK received signal eye diagram for different P_{opt} levels (time scale is 20 ns/div), adopted from [180].

The measured eye-diagrams for the received signals are presented in Figure 7-3. The eye-diagrams clearly show the superior performance results of coherent detection of BPSK modulation compared to the often used OOK coding from the results reported in [180]. The height of the eye-opening is smaller in the presence of turbulence in Figures 7-3(b) and 7.3(c) due to the fog-induced signal intensity fluctuations. In addition, the height of the eye-opening is bigger at the T_x side compared to the R_x. Furthermore, Table 7-2 presents the visibility values corresponding to the measured values of T, the Q-factor, and the corresponding BER for $P_{opt} = 0.6$ dBm. As the amount of fog particles inside the chamber increased, the representing visibility decreased, thus resulting in an increased scattering of the optical beam. The magnitude of the eye height dropped from

3.7 mV without turbulence to 1.2 mV when fog turbulence was injected at the receiver side (R_x), as the visibility also dropped from 850 m to 37 m.

Table 7-2: Experimentally measured values for a 0.6 dBm transmitted signal.

Visibility (m)	Q-Factor	BER	Q-Factor with Lens	BER with Lens
850	4.25	3.24×10^{-6}	4.82	3×10^{-6}
540	4.03	2.79×10^{-05}	4.5	2.03×10^{-05}
224	3.5	1.07×10^{-5}	4	2.7×10^{-4}
165	3.17	2.33×10^{-4}	3.8	2.01×10^{-4}
66	2.09	4.02×10^{-3}	3.5	3.2×10^{-3}
37	1.43	1.83×10^{-02}	2.6	1.01×10^{-02}

The combination of the convex and concave mirrors were used to improve the beam spot size, and therefore, enhance the link performance. The combination of the convex and concave mirrors configuration is illustrated in Figure 7-4. The incident laser beams, which were parallel to the principal axis of the convex mirror with a focal length f' of 1.2 m, were reflected divergently. Since the focus and the reflecting surface are on the opposite side of the convex mirror, the laser beams passed through the focus by extending the reflected ray behind the convex mirror. The divergent laser beams from the convex mirror were then directed on the concave mirror. The focal points of the convex and concave mirrors must be overlapped so that the reflected laser beams from the concave mirror are almost parallel. The optical beam footprints of the reflection spots on the surface of the fixed reflecting mirror are depicted in Figure 7-5. They illustrate the beam spot at every reflection for different fog conditions. Over a link span of 9 m, and with 3 reflections, the spot size increased from 1.6×4.5 cm^2, as per the spot size at the source (Figure 7-5 (a)), to 5×6 cm^2 and 8×7.5 cm^2 (Figures 7-5 (b) and (c) respectively), for no fog, moderate fog and dense fog scenarios respectively, which is due to the beam divergence caused by the fog effect at the receiver side and the non-ideal mirrors.

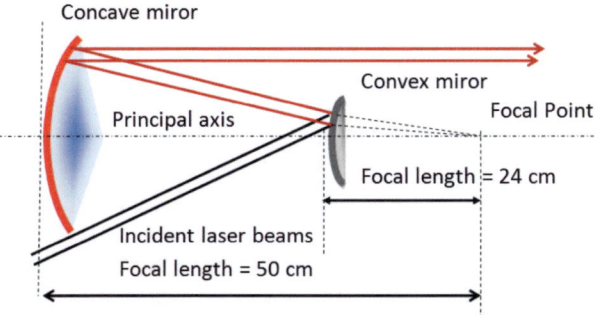

Figure 7-4: Principles of the beam reflections between the convex and concave mirrors.

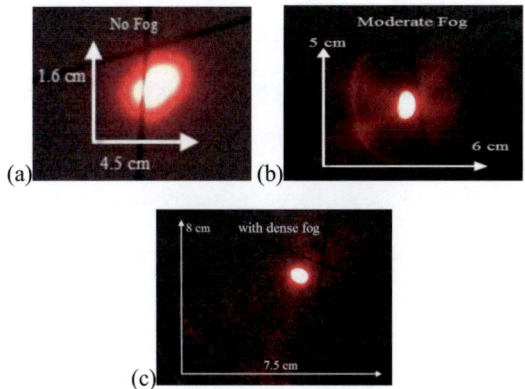

Figure 7-5: Pictures of the reflection spots showing the laser beam divergence at the receiver side Rx with: (a) no fog, (b) moderate fog, and (c) dense fog.

7.3. Performance Analysis of QPSK FSO at Different Wavelengths under Fog Conditions

Aerosol scattering and absorption due to rain, snow, and fog result in significant optical power attenuation, beam spreading and link distance reduction, gravely impairing system performance. Fog, compared to other atmospheric conditions, is the predominant source of attenuation, which reduces the FSO link availability from a few kilometres to less than hundreds of meters [160, 182]. Hence, it is appropriate to investigate whether this challenge strongly affects one wavelength range more than others. Various studies have reported that fog affects the intensity of FSO signals differently depending on the wavelengths of the signals [183, 184, 185]. The studies were carried out applying the OOK modulation scheme, as in [183, 184].

Kim and Korevaar [15] reported that the atmospheric attenuation of laser power is due to the random function of the weather, and can vary from 0.2 dB/km in exceptionally clear weather (i.e. 50-km visibility) to 350 dB/km in very dense fog (i.e. 50-m visibility). The authors evaluated the equation for atmospheric attenuation as a function of wavelength, which is frequently used in free space optics literature [186, 187], for 0.75 µm and 1.55 µm wavelengths. Therefore, in this part of the research, the effect of fog on QPSK FSO links at 0.65 µm, 1.55 µm, and 10 µm wavelengths is investigated in terms of the power losses at the Rx, the link performance in terms of the received power level, and the Q factor under clear and fog-induced atmospheric attenuation.

7.3.1. Mathematical Model

7.3.1.1. Link Budget

The link budget is useful for estimating a communication system's theoretical power-limited link range. The link distance and wavelength, types and efficiencies of transmitters, and power loss sources are key factors that define the FSO link budget performance. The Friis transmission formula [190, 191] provides the received power at the detector side:

$$P_R = P_T \eta_T \eta_R (\lambda/4\pi L)^2 G_T G_R L_T L_R, \qquad (7.8)$$

where P_R is the received signal power, P_T is the transmit signal power, η_T and η_R are the Tx and Rx radiation efficiencies respectively, and G_T and G_R, are the Tx and Rx gain respectively, and are given by:

$$G_T = (\pi D_T / \lambda)^2, \qquad (7.9)$$

$$G_R = (\pi D_R / \lambda)^2, \qquad (7.10)$$

where D_T and D_R are the Tx and Rx apertures respectively, and L_T and L_R are the Tx and Rx pointing-loss factors respectively, based on pointing error angles, and are given by:

$$L_T = \exp(-G_T(\theta_T)^2), \qquad (7.11)$$

$$L_R = \exp(-G_R(\theta_R)^2), \qquad (7.12)$$

where θ_T and θ_R are the Tx and Rx pointing-errors respectively.

7.3.1.2. Atmospheric Attenuation Modelling

Atmospheric attenuation is the process whereby some, or most of the electromagnetic wave energy, is lost by traversing the atmosphere. Therefore, the atmosphere causes signal degradation and attenuation in an FSO system in various ways, including absorption, scattering, and scintillation. The exponential Beers-Lambert law describes the laser power attenuation over an atmospheric channel as [15, 192]:

$$P_R = P_T \exp(-\beta_\lambda L), \qquad (7.13)$$

where β_λ is defined as the atmospheric attenuation coefficient and is expressed as follows [15]:

$$\beta_\lambda = \frac{3.91}{V}(\frac{\lambda}{550nm})^{-q}. \qquad (7.14)$$

According to the definition of attenuation, and from equation (7.13), the total atmospheric loss can be expressed by [15]:

$$\begin{aligned}\alpha_{dB} &= 10\log(\frac{P_T}{P_R}) = 10\log(\frac{1}{\exp(-\beta_\lambda L)}), \\ &= 10\log(\exp(\beta_\lambda L))\end{aligned} \qquad (7.15)$$

Consequently, the atmospheric attenuation in (dB/km) can be given as follows:

$$\alpha_{\frac{dB}{km}} = \frac{\alpha_{dB}}{L} = \frac{[10\log(\exp(\beta_\lambda L))]}{L}. \qquad (7.16)$$

7.3.2. QPSK FSO with Fog

Figure 7-6 shows the block diagram of a terrestrial FSO link. Two FSO transceivers are required to set up a point-to-point communication link. The transmitting and receiving telescopes must be aligned at all times to ensure link operation. We used Optisytem 12 to design and simulate the proposed FSO system based on the parameters presented in Table 7-3.

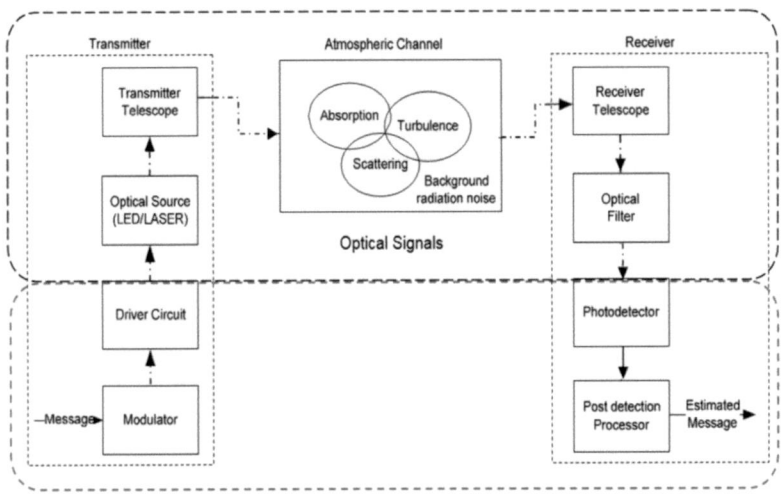

Figure 7-6: Block diagram of a terrestrial FSO link.

Table 7-3: Parameters of the QPSK FSO link simulations.

Parameter		Value
Transmitter	Transmission data rate	1 Gbps
	Link distance S	Up to 5 km
	Transmitted power P_T	20 mW
	Wavelength λ	0.65 μm, 1.55 μm, 10 μm
	Transmitter aperture D_T	10 cm
	Extinction ratio α_e	10 dB
	Beam divergence	2 mrad
	Equipment loss	3.6 dB
	Line width	10 MHz
	Modulation	QPSK
	Transmitter radiation efficiencies η_T	0.85
	Bits per symbol	3
Photodetector	Active area	1 mm^2
	Full angle field of view	150 Deg
	Spectral range	0.65 μm - 10 μm
	Max. wavelength sensitivity	10 μm
	Spectral sensitivity	0.59 A/W

	Rise and fall time	5 ns
	Reversed bias voltage	50 V
	R_x aperture D_R	10 cm
	R_x radiation efficiency η_R	0.75
Lenses	Conical interface	
	Numerical aperture	7mm, 10 mm
	Design wavelength	0.65 μm-10 μm
	Effective focal length f'	4 mm, 11 mm, 19 mm
Receiver	Transamplifier (IC)	AD8015
	Bandwidth	240 MHZ
	Transimpedance amplifier gain	15kΩ

Optical losses are due to the coupling and insertion losses that occur when an optical signal is transmitted or received when using lenses and mirrors. Uncoated glass windows usually attenuate the signal by 4% per surface due to reflections [53]. Imperfect alignment of the FSO transmitter and R_x causes pointing error loss. Typically, these effects are seen for distances in excess of 3 km, at which we might take off an additional dB of power from the link budget. Hence, in this work, we assume no pointing errors. Geometric losses result from beam divergence, as a typical FSO transceiver has an optical beam divergence in the range of 2–10 mrad and 0.05–1.0 mrad (equivalent to a beam spread of 2–10 m, and 5 cm-1 m respectively at 1 km link range) for systems without and with tracking respectively [1]. Geometric losses are usually calculated as the area of the R_x aperture divided by the area of the beam spread over the R_x plane; which is presumed to be perpendicular to the transmitted beam.

According to the mathematical model introduced by equation (7.15), we used MATLAB to regenerate the power loss resulting from both the clear and the fog conditions, as Table 7-4 indicates.

Table 7-4: Atmospheric attenuation in different weather conditions for various wavelengths.

Weather conditions	Visibility (km)	Attenuation (dB/km)		
		0.65 μm	1.55μm	10μm
Clear	12	1.13	0.36	0.03
	23	0.59	0.19	0.01
	40	0.34	0.11	0.009
Fog	0.1	169.80	169.80	169.80
	0.6	27.83	25.51	21.17
	0.8	20.19	15.55	8.89

The effect of fog weather conditions is related to the size distribution of the scattering particles q and V. For clear air and high visibility ($V = 23$ km), the effect of the atmosphere on the signal power levels is nearly negligible for all wavelengths. The situation changes, however, in the case of fog atmospheric conditions and low visibility ($V \leq 0.5$ km). In In the case of moderate and thick fog conditions, attenuation levels increase significantly for all wavelengths, whereas in the case of light fog, the 10μm wavelength achieves significantly less attenuation than the 1.55 μm or 0.65 μm wavelengths. An APD Photodiode, which is widely used in the FSO systems [193], was used as the R_x, a BER of 10^{-9} was adopted as a threshold in all simulations because it is the desired target for all practical FSO link designs [47, 194].

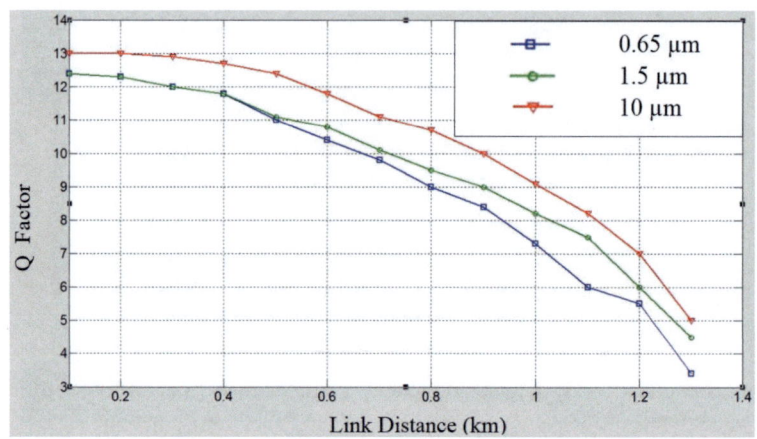

Figure 7-7: Q-Factor vs. link distance (km) at a 2 km visibility limitation factor for 0.65 μm, 1.55 μm, and 10 μm wavelengths.

Figure 7-7 shows that when the link distance is less than 0.5 km, the received signal quality is approximately the same for the 0.65 μm and 1.55 μm wavelengths for a visibility limitation factor of 2 km. However, the Q-factor is higher at 10 μm than at either of the other wavelengths. Additionally, for each wavelength, the quality of the received signal decreases as the link distance increases, and vice versa.

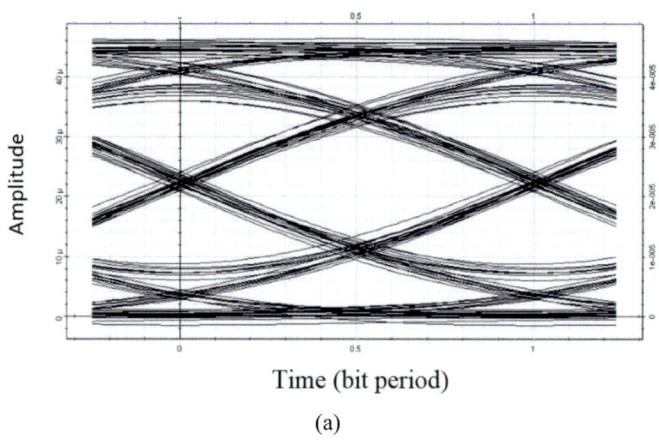

(a)

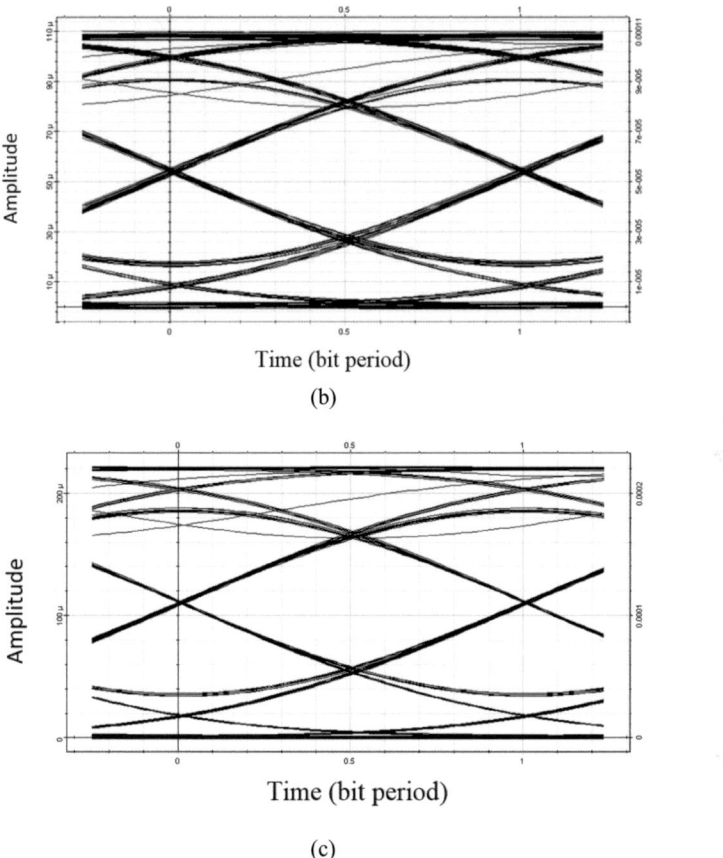

Figure 7-8: The measured eye diagram of the received QPSK signal at the Rx side at: (a) 0.65 μm wavelength, (b) 1.55 μm wavelength, and (c) 10 μm wavelength (x-axis in ns).

Figure 7-8 depicts the measured eye-diagrams of the received signals under light fog attenuation within a visibility of $V = 0.8$ km. The height of the eye opening is smaller at the 0.65 μm and 1.55 μm wavelengths than at the 10 μm wavelength. Although degraded due of the presence of fog along the FSO link, the magnitude of the eye height at 10 μm wavelength is 0.12 mV, the highest, followed by 0.049 mv and 0.019 mv at 1.55 μm and 0.65 μm respectively. Besides applying robust modulation techniques to overcome the fog impairments affecting the FSO link, these findings show that the performance is better at 10 μm wavelength than it is at the shorter wavelengths. However, further optimisation are required to improve the FSO system's link reliability and performance.

7.4. Performance Analysis of Hybrid PPM-BPSK- SIM, PPM and BPSK-SIM FSO with Fog

The performance of an FSO link can be improved by employing an appropriate modulation scheme, and by considering its power and bandwidth efficiencies. The hybrid PPM-BPSK-SIM scheme has been introduced and investigated in a turbulent atmospheric environment [195], indicating its superiority over 2-PPM and BPSK for strongly turbulent atmospheric conditions. However, limited work has been conducted under fog conditions. Thus, we aim to study the performance of FSO links applying the hybrid PPM-BPSK-SIM, which combines the advantage of the PPM and BPSK-SIM, over dense fog atmospheric conditions, and investigate the hybrid scheme's potential to offer improved performance compared to the BPSK-SIM and PPM schemes in a turbulence channel.

Unlike baseband modulation schemes (i.e. PAM) where information is encoded in the amplitude of the optical carrier, the information in BPSK-SIM is encoded in the phase of the RF subcarrier. Hence, BPSK-SIM does not need an adaptive threshold to perform optimally in atmospheric turbulence [196]. PPM is known for its unparalleled power efficiency and does not require an adaptive threshold. However, the PPM scheme suffers from a low spectral efficiency and a high system complexity due to the requirement for both slot and symbol synchronisations.

7.4.1. Characterisation of Fog-induced Attenuation Using A Laboratory Chamber

According to the measurements carried out in [160] in the presence of fog, under high water vapour concentration conditions, the water condenses into tiny water droplets of 1–20 μm radius in the atmosphere. There are different types of ROF, which are categorised on the basis of their formation mechanisms, such as convection fog, advection fog, precipitation fog, valley fog, and steam fog [197]. Steam fog is localised and is created by cold air passing over a much warmer water or moist land [197], [198]. It is possible to simulate this form of fog in the laboratory by achieving H close to 95%. Hence, to mimic ROF, artificial fog can be generated by means of water-based steam. The setup used to assess the performance of the proposed systems is shown in Figures. 7-12(a) and (b). It consisted of an enclosed atmospheric chamber with dimensions of 550×30×30 cm^3 (see Figure 7-12(a)). The maximum link span achieved using the chamber with no reflection was ~ 6 m. The chamber had two compartments, each with a vent to allow air circulation using a fan.

Table 7-5: Parameters of the FSO link with various modulations schemes.

Parameter		Value
Data source	PRBS length	$2^{13} - 1$
	Modulation	BPSK-SIM, 4-PPM, 4-PPM-BPSK
	Data rate	25 Mbps
Laser diode	Peak wavelength	830 nm
	Maximum optical power	10 mW
	Maximum peak to peak voltage	500 mv
	Class	3B
	Beam size at aperture	5 mm × 2 mm
	Beam divergence	<10 mrad

	Modulation bandwidth	50 MHz
Photodetector (PDA36A)	Wavelength at maximum sensitivity	900 nm
	Spectral range of sensitivity	750–1100 nm
	Active area	1 mm^2
	Half angle field of view	± 75 Deg
	Spectral sensitivity	0.60 A/W at 830 nm
	Rise and fall time	5 ns
	Reversed bias voltage	40 V
	Diameter	3.4 cm
	Focal length	20 cm
	Receiver sensitivity	- 32 dBm (at 25Mbps & BER = 10-6)
	Effective focal Length f'	20 cm
Receiver	Transamplifier (IC)	AD8015
	Bandwidth	240 MHz
	Transimpedance amplifier gain	15 kΩ

The R_x front-end consisted of an optical concentration lens and a PIN photodetector integrated with a wide-bandwidth TIA. The photocurrent of the PIN was amplified using the TIA, the output of which was captured using a digital oscilloscope for further offline processing, as presented in Figure 7-12 (c). The fog was generated using a fog machine (water steam) with 100% humidity to replicate the ROF. The chamber had two compartments, each with a vent to allow air circulation using a fan. The fog was injected into the chamber at a rate of 0.97 m^3/sec. The fog intensity within the chamber was controlled by a number of fans and ventilation inlet/outlets allowing us to manage the fog flux among the chamber homogeneously and control the visual contrast (transmittance) of the FSO link.

At the R_x side, the received optical power was converted to an electrical signal by means of an optical receiver (ORx), and a real-time oscilloscope (in our experiment, we used an Agilent DSO80604B infiniium High Performance Oscilloscope) to record the voltages. A virtual instrument script was developed in National Instrument (NI) LabVIEW 2012 [206] to control the devices and record the received data sets, as shown in Figure 7-13. The captured signals were than processed in MATLAB. In order to assess the quality of each signal, two parameters were extracted from the captured signals. Furthermore, in order to ensure a fair comparison between the modulations schemes, the same average optical transmit power P_T was maintained and the Q-factor and BER of the received signal were simultaneously measured for different fog conditions (i.e. from low to high visibilities).

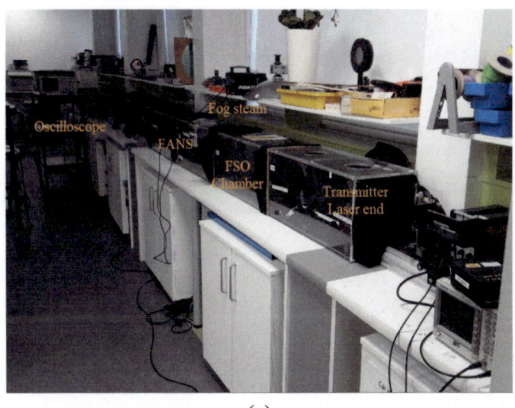

(a)

(b)

121

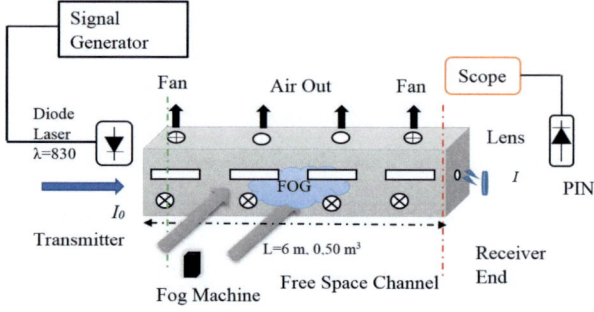

(c)

Figure 7-9: (a) the laboratory controlled atmospheric chamber and FSO link setup, (b) the receiver end of the setup, and (c) a schematic of the fog chamber and FSO link setup.

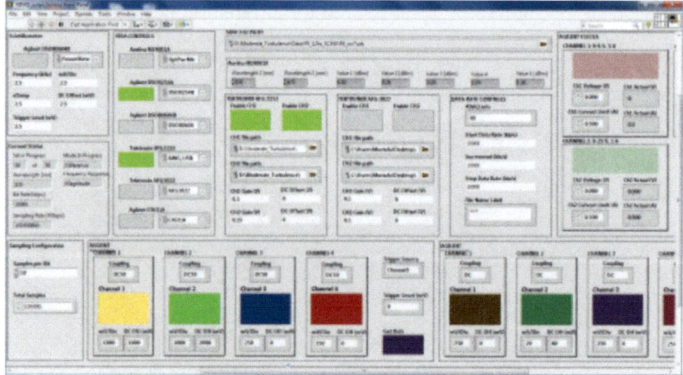

Figure 7-10: A screenshot of the virtual instrument script created in NI LabVIEW.

7.4.2. Hybrid 4-PPM-BPSK-SIM, BPSK-SIM and 4-PPM with Fog

The experiment was carried out by filling the chamber with a controllable amount of fog to achieve a visibility range from very low to high visibility. In the experiment, fog was permitted to settle down homogenously in the chamber before the measurement was taken. The most important impact of the fog is the attenuation of the optical beam by the Mie scattering and absorption. Both phenomena are caused by the interaction of the particles with the electromagnetic waves spreading over the channel. This interaction depends on the characteristic size of obstacle or particle, refractive index and the wavelength of the optical beam. Due to the very small value of the imaginary part of the refractive index, the absorption (Rayleigh scattering) in the IR waveband by fog and aerosols is negligible [197].

The major contributors to the dispersion of the beam are particle sizes, which are close to the optical beam wavelength. Normally, both fog and haze are major contributors to the Mie scattering due to the optical wavelengths being in the order of 1 - 15 μm. Due to the random nature and occurrence of the fog, its impact on FSO links performance can be promptly measured and characterized by the visibility data [178, 198].

Table 7-6: Measured T and visibility values.

Fog	Dense	Thick	Moderate	Light
T	< 0.31	0.31-0.71	0.71-0.8	0.8-0.9
V (m)	< 70	70-250	250-500	500-1000

In this study, we evaluate the scattering coefficient β_λ corresponding to the measured T at 570 nm using equation (7.5). The visibility V is evaluated using (7.4) with the wavelength independent model, known as Kim model, for ranges < 500 m and for different transmittance values (see Table 7-5). The Q-factor results depicted in Figure 7-14 show that the Hybrid 4-PPM-BPSK-SIM and 4-PPM modulation signalling formats are more robust to fog impairments on the FSO link than BPSK-SIM. However, the behaviour of the three modulations schemes under fog conditions are similar in absolute terms at T = 1, where the Q-factor is 8.5 for 4-PPM, 6.5 for Hybrid 4-PPM-BPSK-SIM and 6.2 for BPSK-SIM. On the other hand, the Hybrid 4-PPM-BPSK and the 4-PPM schemes exceed that values for T > 0.2. Figure 7-14 illustrates the predominant performance gain of the Hybrid 4-PPM-BPSK-SIM format over the rest of modulation schemes in moderate, thick and dense fog conditions.

Furthermore, Table 7-7 presents the visibility values and the corresponding measured values of T and the BER for P_{opt} = 0.60 dBm. As the amount of fog particles inside the chamber increases, the corresponding visibility decreases, thus resulting in increased scattering of the optical beam. This, in turn, causes the distribution of bits '1' and '0' to be more flat due to the reduction of the height of the eye diagram.

Table 7-7: The experimentally measured values for the 0.6dBm transmitted signal BER.

Visibility (m)	BER Hybrid 4-PPM-BPSK-SIM	BER 4-PPM	BER BPSK-SIM
450	9.479 x 10^{-17}	4.016 x 10^{-10}	9.86 x 10^{-9}
320	4.016 x10^{-10}	1.898 x 10^{-7}	3.33 x 10^{-7}
140	2 x 10^{-3}	13.4 x 10^{-3}	25 x 10^{-3}
100	10.7 x 10^{-4}	81.9 x 10^{-3}	22.7 x 10^{-2}
65	62 x 10^{-3}	66.8 x 10^{-2}	136 x 10^{-2}

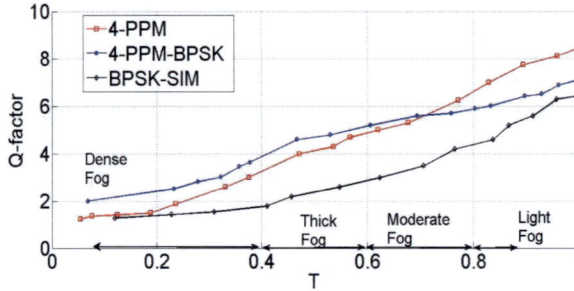

Figure 7-11: The measured Q-factor values for the Hybrid 4-PPM-BPSK, 4-PPM and BPSK-SIM received signals at the same P_T and 5Mbit/s data rate for different T link values and fog conditions.

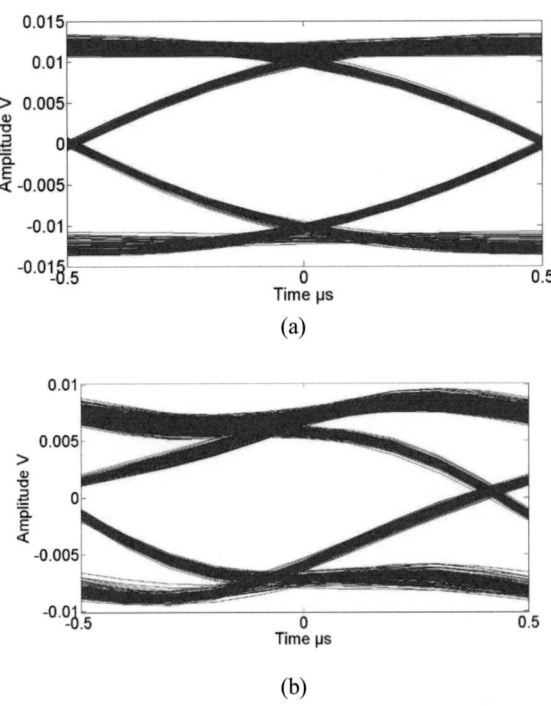

Figure 7-12: The measured eye diagrams of the received Hybrid 4-PPM-BPSK-SIM signal at a data rate of 25 Mb/s with: (a) no fog and $V = 320\ m$, and (b) fog and $V = 100$ m.

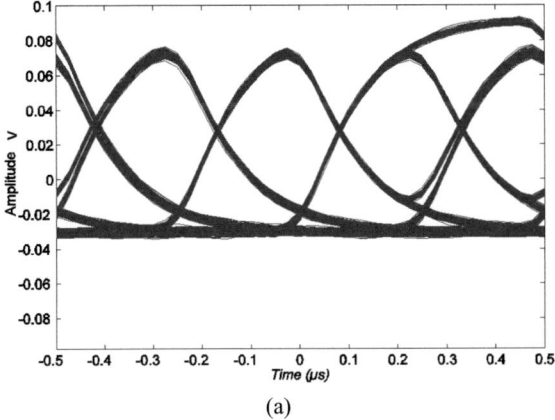

(a)

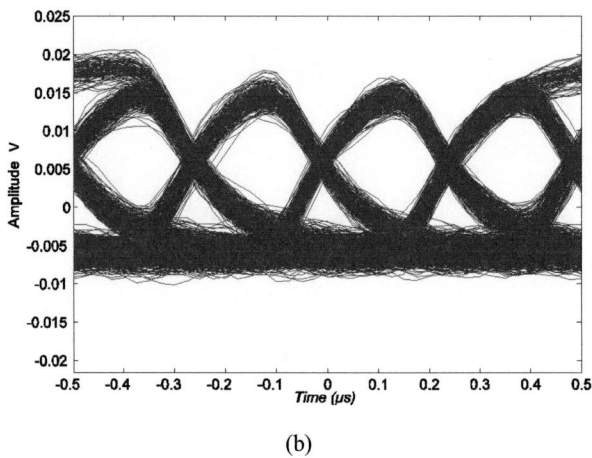

(b)

Figure 7-13: The measured eye diagrams of the received 4-PPM signal at a data rate of 25 Mb/s with: (a) no fog and $V = 320$ m, and (b) fog and $V = 100$ m.

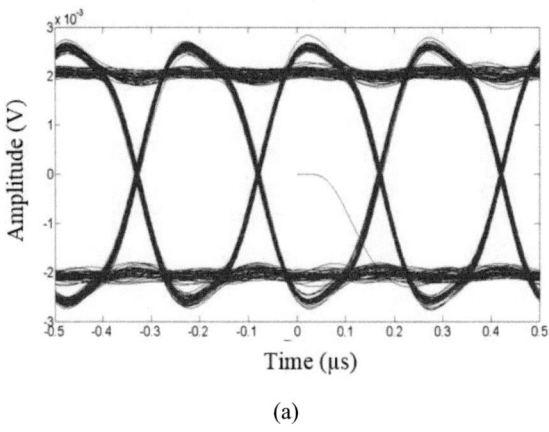

(a)

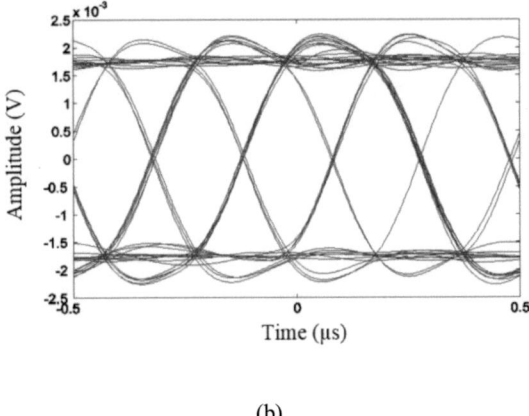

(b)

Figure 7-14: The measured eye diagrams of the received BPSK-SIM signal at a data rate of 25 Mb/s with: (a) no fog, $V = 320$ m, and (b) fog and $V = 100$ m.

The eye-opening is bigger in the presence of fog-induced attenuation for the Hybrid 4-PPM-BPSK-SIM signal compared to the 4-PPM and BPSK-SIM signals. On the other hand, the eye-opening is bigger for the 4-PPM signal than the BPSK-SIM signal, indicating that the Hybrid 4-PPM-BPSK is less sensitive to intensity fluctuations under fog conditions.

7.5. Summary

This chapter described the possible methods of mitigating the effects of fog on FSO links by implementing different modulation schemes such as BPSK-SIM, 4-PPM, Hybrid 4-PPM-BPSK-SIM and QPSK in a laboratory controlled fog environment. The performance of PPM_BPSK, 4-PPM and BPSK-SIM for $N = 1$ were investigated and compared. The effects of a wide range of visibility values on the BER performance of FSO links in the presence of fog was also measured and investigated. The obtained results indicated the dependency of the fog intensity variation on the performance of the FSO link. Furthermore, the hybrid 4-PPM-BPSK-SIM signalling format demonstrated improved resistance to thick fog impairments compared with the 4-PPM and BPSK-SIM schemes. In addition, the BER characteristics of the three modulations schemes over fog affected channel were studied. The results showed that the use of the hybrid 4-PPM-BPSK-SIM modulation format improves the performance of the FSO system over a thick fog channel. Moreover, the influence of wavelength on FSO performance was investigated. The shorter wavelengths was found to be more limited in laser power than the longer wavelengths. Furthermore, the fog-induced attenuations were shown to have a reduced impact on longer wavelengths. In light of these results, it is clear that opting for an appropriate modulation technique that considers the system's attenuation level is not enough to mitigate fog-induced impairment.

8. CONCLUSIONS AND FUTURE WORKS

The requirement for FSO-based back-up and complementary links to RF technology, especially for the "last mile" in access networks, has increased substantially. This is due to a number of advantageous features such as an un-regulated and license free transmission bandwidth spectrum, a high data rate transmission, a low power consumption, an enhanced security, as well as immunity to electromagnetic interference. However, the challenge imposed by the atmospheric channel in the form of turbulence, smoke and fog, greatly diminishes the system performance and the availability of FSO links. Detailed modelling the FSO channel is an important task that needs to be undertaken to improve the performance of FSO links. To carry out the objectives of this research, the fundamental principles of FSO technology and its sub-blocks were reviewed and discussed in Chapter 2. The applications and features of FSO technology that make it more practical than the existing RF technologies, were detailed.

Following this, a discussion on the atmospheric channel, both in terms of atmospheric attenuation and turbulence, was presented. Chapter 3 highlighted and discussed the three main reported models for irradiance fluctuation in atmospheric turbulence, which are important in predicting the reliability of an optical system operating in such an environment. The lognormal model is mathematically tractable, however only valid in the weak turbulence regime. Beside the weak turbulence regime, the Gamma-Gamma model is more suitable but lacks the mathematical convenience. The negative exponential model is only employed in the saturation regime. A number of empirical fog models were considered. Kim and Naboulsi models were determined to be the best models in estimating the attenuation of the optical signal induced by fog for the wavelength range of 0.5 μm to 0.9 μm, when $V = 1$ km.

In accordance with the research objectives, this book focused on optimising the system performance during atmospheric turbulence as well as in the presence of fog, through the use of several modulation techniques such as BPSK-SIM, 4-PPM, Hybrid 4-PPM-BPSK, and QPSK. In order to overcome the performance limitation of OOK-FSO in turbulent atmospheric channels, the BPSK-SIM, 4-PPM and hybrid 4-PPM-BPSK-SIM modulation schemes were proposed and their error performance analysis under atmospheric turbulence was presented in Chapter 6. The results showed that PPM-BPSK offered similar performance to the 2-PPM, a superior performance to BPSK-SIM while having the same bandwidth, but an inferior performance to 4-PPM for all turbulence levels. Furthermore, the experimental investigation indicated that the 4-PPM technique is more robust to turbulence induced impairments in the FSO link than the BPSK-SIM technique.

Fog, as a dominant source for power attenuation and potentially reducing FSO links' availability, was discussed in Chapter 7. An experimental evaluation of the performance of BPSK with coherent detection, 4-PPM and Hybrid 4-PPM-BPSK under the effect of fog for FSO communication links was carried out in a controlled laboratory test-bed. The effects of low to high visibility on the FSO link BER performance in the presence of fog were also measured and investigated. Furthermore, a performance analysis of FSO systems using QPSK modulation techniques over a fog-affected atmospheric channel was provided for various wavelengths of operation. The shorter wavelengths were found to be more limited in laser power output than the longer wavelengths. Furthermore, the results showed that fog-induced attenuation has less impact at longer wavelengths. As such, opting for an appropriate modulation technique that considers the system's attenuation level is not enough to mitigate fog-induced impairments.

Finally, an experimental evaluation of the performance of the Hybrid 4-PPM-BPSK-SIM, 4-PPM and BPSK-SIM modulation schemes was conducted under the effect of fog in a controlled laboratory test-bed. The hybrid 4-PPM-BPSK-SIM signalling format offers improved resistance to thick fog impairment compared to the 4-PPM and BPSK-SIM schemes.

Whilst the objectives listed in Chapter 1 of this book have been achieved, this work was never envisaged to be comprehensive to the degree of addressing the entire optical wireless communication research challenges. Since the amount of research scope and time expected to do so is beyond the scope of this book, recommendations of further research work, which can be carried out to broaden the work covered here, are given below.

Hybrid FSO/RF communication employing channel coding:

Weather, propagation distance, scattering, absorption, turbulence, pointing error effects, laser wavelength, and data rate are some of the key elements that impact the overall performance of FSO link. The main challenge facing FSO technology is the attainment of 99.999% link availability during all weather conditions [199]. A hybrid FSO/RF link combined with channel coding is one possible option to address this challenge. In such a scheme, the RF system could be employed as the back-up link, though at a reduced data rate while the fog/turbulence levels are moderate to high, and the channel coding can also improve the overall system dependability [200, 201].

Radio over FSO:

Transmission of modulated RF signals using FSO communication links has been widely researched recently [204]. The radio over free-space optics (RoFSO) communications system has the potential to offer a cost effective and a reliable technology for bridging advanced wireless technologies network facilities. RoFSO is a next generation access technology suitable for transmission of heterogeneous wireless service signals, particularly in areas with limited broadband connectivity. There is a need to conduct initial investigations focusing on examining the deployment environment features, which influence the performance of RoFSO systems [205].

Spatial Diversity:

FSO signals are severely impaired in strong turbulence environments, which can be mitigated by spatial diversity techniques such as MIMO antennas that can be equipped on the transmitter and/or receiver sides [202, 203].

Multipath Diversity:

Wireless networks suffer signal fading due to multipath propagation, which can be mitigated by time-, frequency- or spatial diversity techniques. With regard to spatial diversity, multiple antennas can be equipped on the transmitters and/or receivers; which is difficult to implement on a sensor node or a mobile terminal due to size limitations and hardware complexity.

Adaptive Optics:

Although adaptive optics technology was investigated in this research, it is considered to be a promising mitigation technology of atmospheric attenuation. Therefore, it is recommended to extend applying adaptive optics for future FSO system. Thanks to adaptive optics, it is theoretically possible to pre-compensate the effects of atmospheric turbulence on the laser beam [214]. Several solutions can be considered, which can be classified into two families: pre-compensation of the phase only [211], or of the phase and amplitude (i.e. a full-wave correction) [212]. In a research conducted by Schwartz et al. [213], the two methods were compared through simulations, and the results indicated that for long-range applications, full-wave correction performs significantly better.

9. REFERENCES

[1] W. O. Popoola, "Subcarrier intensity modulated free-space optical communication systems," *in School of Computing, Engineering and Information Sciences. Doctor of Philosophy Newcastle: University of Northumbria*, September 2009.

[2] M. S. Awan, P. Brandl, E. Leitgeb, F. Nadeem, T. Plank, and C. Capsoni, "Results of an optical wireless ground link experiment in continental fog and dry snow conditions," *10th International Conference on Telecommunications*, pp. 45-49, June 2009.

[3] A. Hashmi, A. Eftekhar, S. Yegnanarayanan, and A. Adibi, "Analysis of optimum adaptive optics systems for hybrid RF-wireless optical communication for maximum efficiency and reliability," *4th International Conference on Emerging Technologies*, pp. 62-67, 19 October 2008.

[4] H. Hemmati, "Interplanetary laser communications," *Optics and Photonics News*, vol. 18, pp. 22-27, November 2007.

[5] S. Zoran, F. Bernhard, and L. Hanspeter, "Free-space laser communication activities in Europe: SILEX and beyond," *IEEE Lasers and Electro-Optics Society (LEOS) 19th Annual Meeting*, pp. 78-79, October 2006.

[7] F. E. Goodwin, "A review of operational laser communication systems," *Proceedings of the IEEE*, vol. 58, pp. 1746-1752, 1970.

[8] S. C. Gupta, Optoelectronic Devices and Systems: *Prentice-Hall of India Pvt. Limited*, 2005.

[9] D. J. Geisler et al., "Multi-aperture digital coherent combining for free space optical communication receivers," *Opt. Express*, vol. 24, pp. 12661–12671. 2016.

[10] J. C. Juarez, A. Dwivedi, A. Mammons, S. D. Jones, V. Weerackody, and R. A. Nichols, "Free-space optical communications for next-generation military networks," *Communications Magazine IEEE*, vol. 44, pp. 46-51, 2006.

[11] H. Tapse, D. K. Borah, and J. Perez-Ramirez, "Hybrid optical/RF channel performance analysis for Turbo codes," *IEEE Transactions on Communications*, vol. 59, pp. 1389-1399, 2011.

[12] N. D. Chatzidiamantis and G. K. Karagiannidis, "On the distribution of the sum of gamma-gamma variates and applications in RF and optical wireless communications," *IEEE Transactions on Communications*, vol. 59, pp. 1298-1308, 2011.

[13] M. Gregory and S. Badri-Hoeher, "Characterization of maritime RF/FSO channel," *International Conference on Space Optical Systems and Applications (ICSOS)*, pp. 21-27, 2011.

[14] P. Y. Zhou and F. Khan, "An introduction to millimeter-wave mobile broadband systems," *IEEE Communications Magazine*, vol. 49, pp. 101-107, June 2011.

[15] I. Isaac. Kim, B. McArthur, and E. Korevaar, "Comparison of laser beam propagation at 785 nm and 1550 nm in fog and haze for optical wireless communications," *SPIE Proceedings: Optical Wireless Communications III*, vol. 4214, pp. 26-37, 2001.

[16] M. Bettayeb and S. F. A. Shah, "State of the art ultra-wideband technology for communication systems: a review," *Proceedings of the 10th IEEE International Conference on Electronics, Circuits and Systems, (ICECS)*, vol. 3, pp. 1276-1279, 2003.

[17] J. N. Pelton, "Global satellite communications technology and systems," *International Technology Research Institute, World Technology (WTEC) Division*, Baltimore, 1998.

[18] E. Leitgeb, M. S. Awan, P. Brandl, T. Plank, C. Capsoni, R. Nebuloni, T. Javornik, G. Kandus, S. S. Muhammad, and F. Ghassemlooy, "Current optical technologies for wireless access," *in Telecommunications,*

ConTEL, 10th International Conference on, pp. 7-17, 2009.

[19] S. Ghosh, K. Basu, and S. K. Das, "An architecture for next-generation radio access networks," *Network, IEEE*, vol. 19, pp. 35-42, 2005.

[20] W. Gappmair and M. Flohberger, "Error performance of coded FSO links in turbulent atmosphere modeled by gamma-gamma distributions," *IEEE Transactions on Wireless Communications*, vol. 8, pp. 2209-2213, 2009.

[21] H. E. Nistazakis, T. A. Tsiftsis, and G. S. Tombras, "Performance analysis of free-space optical communication systems over atmospheric turbulence channels," *Communications, IET*, vol. 3, pp. 1402-1409, 2009.

[22] S. A. Zabidi, W. A. Khateeb, M. R. Islam, and A. W. Naji, "The effect of weather on free space optics communication (FSO) under tropical weather conditions and a proposed setup for measurement," *in Computer and Communication Engineering (ICCCE), International Conference on*, pp. 1-5, 2010.

[23] K. W. Fischer, M. R. Witiw, and J. A. Baars, "Atmospheric Laser Communication," *Bulletin of the American Meteorological Society*, vol. 85, pp. 725-732, 2004.

[24] J. P. Perera, "A Low-Cost, Man-Portable, Free-Space-Optics Communications Device for Ethernet Applications," *DTIC Document*, 2004.

[25] V. Ramasarma, "Free space optics: A viable last-mile solution," *Bechtel Telecommunications Technical Journal*, vol. 1, pp. 22-30, 2002.

[26] C.-H. Cheng, "Signal processing for optical communication," *IEEE Signal Processing Magazine*, pp. 88-94, Jan. 2006.

[27] H. Willebrand and B. S. Ghuman, Free Space Optics: Enabling Optical Connectivity in today's network. Indianapolis: *SAMS publishing*, 2002.

[28] T.-Y. Lu and W.-Z. Chen, "A 3–10 GHz, 14 bands CMOS frequency synthesizer with spurs reduction for MB-OFDM UWB system," *IEEE Transactions on Very Large Scale Integration (VLSI) Systems*, vol. 20, pp. 1-11, 2011.

[29] Z. Yin, S. Wu, Z. Shi, Z.Wu "A novel waveform with notched frequency band UWB communication system," Radar (RADAR), *CIE International Conference on*, pp. 1-3, 2016.

[30] R. C. Qiu, H. Liu, and X. Shen, "Ultra-wideband for multiple access communications," *IEEE Communications Magazine*, vol. 43, pp. 80-87, 2005.

[31] S. K. Das and K. Basu, "Emergent technology based radio access network (RAN) design framework for next generation broadband wireless systems," *First International Conference on Broadband Networks*, 2004.

[32] J. Tóth; Ľ. Ovseník; J. Turán, "Challenges in modern wireless optical communication systems — Free space optics: Data flow control in FSO monitoring system," *Carpathian Control Conference (ICCC), 17th International* pp. 753 - 756, June 2016.

[33] W. O. Popoola, Z. Ghassemlooy, J. I. H. Allen, E. Leitgeb, and S. Gao, "Free- space optical communication employing subcarrier modulation and spatial diversity in atmospheric turbulence channel," *Journal of Optoelectronics, IET*, vol. 2, pp. 16-23, 2008.

[34] S. M. Navidpour, M. Uysal, and M. Kavehrad, "BER performance of free space optical transmission with spatial diversity," *Wireless Communications, IEEE Transactions on*, vol. 6, pp. 2813-2819, 2007.

[35] D. Kedar and S. Arnon, "Urban optical wireless communication networks: the main challenges and possible solutions," *IEEE Communications Magazine*, vol. 42, pp. S2-S7, May 2004.

[36] S. Karp, R. M. Gagliardi, S. E. Moran, and L. B. Stotts, Optical Channels: fibers, clouds, water and the atmosphere. *New York: Plenum Press*, 1988.

[37] I. I. Kim, J. Koontz, H. Hakakha, P. Adhikari, R. Stieger, C. Moursund M.Barclay, A. Stanford, R. Ruigrok, J. J. Schuster, and E. J. Korevaar, "Measurement of scintillation and link margin for the TerraLink laser communication system," Wireless Technology and Systems: *Millimeter Wave and Optical, Proceedings of SPIE*, vol. 3232, pp. 100-118, 1997.

[38] M. Gebhart, E. Leitgeb, S. S. Muhammad, B. Flecker, C. Chlestil, M. Al

Naboulsi, F. de Fornel, and H. Sizun, "Measurement of light attenuation in dense fog conditions for FSO applications," *in Optics & Photonics*, pp. 58910-12, 2005.

[39] G. Xu, X. Zhang, J. Wei, and X. Fu, "Influence of atmospheric turbulence on FSO link performance," *in Proceedings of SPIE*, pp. 816-823, 2003.

[40] N. Cvijetic, S. G. Wilson, and M. Brandt-Pearce, "Performance bounds for free-space optical MIMO systems with APD receivers in atmospheric turbulence," *Selected Areas in Communications, IEEE Journal on*, vol. 26, pp. 3-12, 2008.

[41] E. Bayaki, R. Schober, and R. K. Mallik, "Performance analysis of MIMO free-space optical systems in gamma-gamma fading, "*Communications, IEEE Transactions on*, vol. 57, pp. 3415-3424, 2009.

[42] H. Yuksel, S. Milner, and C. Davis, "Aperture averaging for optimizing receiver design and system performance on free-space optical communication links," *Journal of Optical Networking*, vol. 4, pp. 462-475, 2005.

[43] Z. Wang, W.-D. Zhong, S. Fu, and C. Lin, "Performance comparison of different modulation formats over free-space optical (FSO) turbulence links with space diversity reception technique," *Photonics Journal, IEEE*, vol. 1, pp. 277-285, 2009.

[44] X. Tang, Z. Ghassemlooy, W. O. Popoola, and C. Lee, "Coherent polarization shift keying modulated free space optical links over a Gamma Gamma turbulence channel," *American Journal of Engineering and Applied Sciences*, vol. 4, pp. 520-530, 2011.

[45] F. Nadeem, B. Flecker, E. Leitgeb, M. Khan, M. Awan, and T. Javornik, "Comparing the fog effects on hybrid network using optical wireless and GHz links, in "*Communication Systems, Networks and Digital Signal Processing*, CNSDSP, *6th International Symposium on*, pp. 278- 282, 2008.

[46] V. Kvicera, M. Grabner, and O. Fiser, "Visibility and attenuation due to hydrometeors at 850 nm measured on an 850 m path," *in Communication*

Systems, Networks and Digital Signal Processing, CNSDSP, 6th International Symposium on, 2008, pp. 270-272, 2008.

[47] M. Uysal, J. T. Li, and M. Yu, "Error rate performance analysis of coded free space optical links over gamma-gamma atmospheric turbulence channels," *IEEE Transactions on Wireless Communications*, vol. 5, pp. 1229–1233, 2006.

[48] Kuo-Nan Liou, "An Introduction to Atmospheric Radiation", *International Geophysics Series*, Volume 84, Second Edition, pp. 90-93, 2002.

[49] Muhammad Ijaz, "Experimental Characterisation and Modelling of Atmospheric Fog and Turbulence in FSO", *in School of Computing, Engineering and Information Sciences, Doctor of Philosophy Newcastle: University of Northumbria*, 2013.

[50] Anil J. Kshatriya, Y. B. Acharya, A. K. Aggarwal, A. K. Majumdar, "Communication performance of free space optical link using wavelength diversity in strong atmospher-ic turbulence", *on Journal of Optics*, vol 44, pp.215-219, 2015.

[51] X. Zhu and J. M. Kahn, "Performance bounds for coded free-space optical communications through atmospheric turbulence channels," *IEEE Transaction on Communications*, vol. 51, pp. 1233-1239, 2003.

[52] X. Zhu and J. M. Kahn, "Mitigation of turbulence-induced scintillation noise in free space optical links using temporal-domain detection techniques," *IEEE Photonics Technology Letters*, vol. 15, pp. 623-625, 2003.

[53] S. Bloom, E. Korevaar, J. Schuster, and H. Willebrand, "Understanding the performance of free-space optics," *Journal of Optical Networking*, vol. 2, pp. 178-200, 2003.

[54] J. T. Li and M. Uysal, "Optical wireless communications: System model, capacity and coding," *Vehicular Technology Conference*, vol. 1, pp. 168-172, 2003.

[55] I. B. Djordjevic, B. Vasic, and M. A. Neifeld, "LDPC coded OFDM over

the atmospheric turbulence channel," *Optical Express*, vol. 15, pp. 6336-6350, 2007.

[56] H. Yamamoto and T. Ohtsuki, "Atmospheric optical subcarrier modulation systems using space-time block code," *in IEEE Global Telecommunications Conference*, (GLOBECOM '03) vol. 6, New York, pp. 3326-3330, 2003.

[57] E. J. Lee and V. W. S. Chan, "Optical communications over the clear turbulent atmospheric channel using diversity," *IEEE Journal on Selected Areas in Communications*, vol. 22, pp. 1896-1906, 2004.

[58] A. Douik, H. Dahrouj, T. Y. Al-Naffouri, M.S Alouini, "Hybrid Radio/Free-Space Optical Design for Next Generation Backhaul Systems," *IEEE Transactions on Communications*, vol. 64, pp. 2563 - 2577, 2016.

[59] K. Prabu, et al., "Spectrum analysis of radio over free space optical communications systems through different channel models," *Optik - International Journal for Light and Electron Optics*, vol. 126, pp. 1142-1145, 2015.

[60] M. A. Esmail; H. Fathallah; M.S. Alouini, "Outdoor FSO Communications under Fog: Attenuation Modeling and Performance Evaluation," *IEEE Photonics Journal*, vol.8, pp. 1-23, 2016.

[61] P. Gupta, P.R. Kumar, "The capacity of wireless networks", *IEEE Trans. Information Theory*. 46, 388-404, 2000.

[62] S. Trisno;" Design and Analysis of Advanced Free Space Optical Communication Systems", *in Department of Electrical and Computer Engineering, Doctor of Philosophy, University of Maryland*, pp. 3-20, 2006.

[63] L.Yang, X. Song, J. Cheng, and J.F. Holzman, "Free-Space Optical Communications Over Lognormal Fading Channels Using OOK With Finite Extinction Ratios", *IEEE Access* 4:1-1, pp. 574-584, 2016.

[64] L.D. Truong, S.T. Luong, L.P. Dinh, H.T. T. Pham and N.T. Dang, "Design and optimization of FSO mesh networks over atmospheric

turbulence and misalignment fading channels," *International Conference on Advanced Technologies for Communications (ATC)*, pp. 356 - 361, 2016.

[65] D. Killinger, "Free space optics for laser communication through the air," *Optics & Photonics News*, vol. 13, pp. 36-42, 2002.

[66] Abdulsalam Ghalib Alkholidi and Khaleel Saeed Altowij, "Free Space Optical Communications—Theory and Practices," *Contemporary Issues in Wireless Communications, Dr. Mutamed Khatib (Ed.), InTech*, 2014. DOI: 10.5772/58884. Available on https://www.intechopen.com/books/contemporary-issues-in-wireless-communications/free-space-optical communications-theory-and-practices, last visited August 2016.

[67] K. Biesecker, "The promise of broadband wireless," *IT Professional*, vol. 2, pp. 31-39, 2000.

[68] G. Hörack, P. Pezzei, E. Leitgeb, M. Tischlinger., "Economic aspects of free space optics as an alternative broadband network connection for the last mile," *Communication Systems, Networks and Digital Signal Processing (CSNDSP), 10th International Symposium on*, pp. 1-5, September 2017.

[69] T. Plank, M. Czaputa, E. Leitgeb, S. S. Muhammad, N. Djaja, B. Hillbrand, P. Mandl, and M. Schonhuber, "Wavelength selection on FSO-links," *Proceedings of the 5th European Conference on Antennas and Propagation (EUCAP)*, pp. 2508-2512, 2011.

[70] J.Tóth, L. Ovseník and J. Turán, "Advanced wireless communication systems —Free space optics: Atmosphere monitoring proposal (Fog and Visibility)," *IEEE 13th International Scientific Conference on Informatics*, pp. 281- 285, 2015.

[71] Y. Li, N. Pappas, V. Angelakis, M.Pióro and D.Yuan, "Optimization of Free Space Optical Wireless Network for Cellular Backhauling," *IEEE Journal on Selected Areas in Communications*, vol. 33, pp. 1841-1854, 2015.

[72] G. Wu, Y. A. Zhang, X. G. Yuan, J. N. Zhang, M. L. Zhang, and Y. P. Li,

"Design and Realization of 10Gbps DPSK System for Free Space Optical Communication," *Applied Mechanics and Materials*, vol. 263, pp. 1150-1155, 2013.

[73] A. Sharma and R. Kaler, "Designing of high-speed interbuilding connectivity by free space optical link with radio frequency backup," *IET Communications*, vol. 6, pp. 2568-2574, 2012.

[74] A. K. Majumdar and J. C. Ricklin, Free-space laser communications: *principles and advances*, New York: NY: Springer, 2008.

[75] T. Yamashita, M. Morita, M. Shimizu, D. Eto, K. Shiratama, and S. Murata, "The new tracking control system for free-space optical communications," *International Conference on Space Optical Systems and Applications (ICSOS)*, pp. 122-131, 2011.

[76] E. Dadrasnia, S. Ebrahimzadeh, and F. R. M. Adikan, "Influence of short range free Space optical atmospheric attenuation in modulated radio signal," *The 2nd International Conference on Computer and Automation Engineering (ICCAE)*, vol. 5, pp. 569-571, 2010.

[77] A. Jendeya, M. El-Absi, N.Zarifeh, S.Ikki, T.Kaiser, "Interference Alignment and Free-Space Optics Based Backhaul Networks," *SCC*; 11th International ITG Conference on Systems, *Communications and Coding*, pp. 1-5, 2017.

[78] A. McCuaig, "Free-space optics: LEDs vs lasers," *in Lightwaveonline*: Penn-Well, 2009.

[79] G. Keiser, Optical Communications Essentials *1st ed. New York: McGraw Hill Professional*, 2003.

[80] O. Bouchet, H. Sizun, C. Boisrobert, F. De Fornel, and P. Favennec, Free Space Optics: *Propagation and Communication. London: ISTE Ltd*, 2006.

[81] N. D. Chatzidiamantis, A. S. Lioumpas, G. K. Karagiannidis, and S. Arnon, "Optical Wireless communications with adaptive subcarrier PSK intensity modulation," *IEEE Global Telecommunications Conference GLOBECOM*, pp. 1-6, 2010.

[82]   Y. Zhao, X. Yu, X. Zheng, I. T. Monroy, and H. Zhang, "Generalized tensor analysis Model for multi-subcarrier analog optical systems," *Journal of Lightwave Technology*, vol. 29, pp. 3144 - 3155, 2011.

[83]   AFGL, "AFGL Atmospheric Constituent Profiles (0–120 km)," AFGL-TR 86-0110, *Air Force Geophysics Laboratory, Hanscom Air Force Base*, Massachusetts, 1986.

[84]   S. G. Narasimhan and S. K. Nayar, "Vision and the atmosphere," *International Journal of Computer Vision*, Vol 48, pp. 233–254, 2002.

[85]   L. Peng, X. Wu, K. Wakamori, T. D. Pham, M. S. Alam, and M. Matsumoto, "Bit error rate performance analysis of optical CDMA time-diversity links over gamma-gamma atmospheric turbulence channels," *IEEE Wireless Communications and Net-working Conference (WCNC)*, pp. 1932-1936, 2011.

[86]   W. K. Pratt, Laser Communication Systems, 1st ed. New York: John *Wiley & Sons, Inc.*, 1969.

[87]   Hemani Kaushal, V.K. Jain, Subrat Kar, "Free Space Optical Communication," *Springer*, Vol.1, 2017.

[88]   X. Tang, Z. Ghassemlooy, S. Kaushal, W. Popoola, M. Uysal, and D.Wu, "Experimental demonstration of polarisation shift keying in the free space optical turbulence channel," *Communications in China Workshops (ICCC), 1st IEEE International Conference on*, pp. 31-36, 2012.

[89]   S. Betti, G. De marchis, and E. Iannone, Coherent Optical Communication Systems, *1st ed. Canada: John Wiley and Sons Inc.*, 1995.

[90]   S. Hranilovic, Wireless Optical Communication Systems vol. 1st. Boston: *Springer*, 2005.

[91]   IEC, "Safety of laser products - Part 1: Equipment classification and requirements," *2nded: International Electrotechnical Commission*, 2007.

[92]   D. J. T. Heatley, D. R. Wisely, I. Neild, and P. Cochrane, "Optical wireless: The story so far," *IEEE Communications Magazine*, vol. 36, pp. 72-82, 1998.

[93] K. Schröder, "Handbook on Industrial laser safety": *Technical University of Vienna*, 2000.

[94] R. Pernice1, A. Andò1, M. Cardinale1, L. Curcio1, S. Stivala1, A. Parisi1, A. C. Busacca1, Z. Ghassemlooy and J. Perez, "Indoor free space optics link under the weak turbulence regime: measurements and model validation", *in IET Communications*, vol. 9, Issue: 1, pp. 62 – 70, 2015.

[95] J. Zeller and T. Manzur, "Free-space optical communication at 1.55 μm and turbulence measurements in the evaporation layer," *SPIE Security & Defence*, Edinburgh, pp. 85400C, 2012.

[96] N. Aida Mohd Nor, E.Fabiyi, M. Mansour Abadi, X.Tang, Z. Ghassemlooy, A. Burton, "Investigation of Moderate-to-Strong Turbulence Effects on Free Space Optics-A Laboratory Demonstration", *The 13th International Conference on Telecommunications, ConTEL*, pp.1-5, Graz, 2015.

[97] N. A M. Nor, J. Bohata, Z. Ghassemlooy, S. Zvanovec, P. Pesek, M. Komanec, J. Libich and M.A. Khalighi, "10 Gbps all-optical relay-assisted FSO system over a turbulence channel," *4th International Workshop on Optical Wireless Communications (IWOW)*, pp. 69-72, 2015.

[98] G. R. Osche, Optical detection theory for laser applications: *Wiley, New Jersey*, 2002.

[99] L. C. Andrews and R. L. Phillips, Laser beam propagation through random media. *Bellingham*: *WA*: *SPIE*, 1998.

[100] L. C. Andrews, R. L. Phillips, and C. Y. Hopen, Laser beam scintillation with applications. *Beingham*: *SPIE*, 2001.

[101] X. Zhu and J. M. Kahn, "Free-space optical communication through atmospheric turbulence channels," *IEEE Transactions on Communications*, vol. 50, pp. 1293-1300, 2002.

[102] S. F. Clifford, "The classical theory of wave propagation in a turbulent medium," in Laser Beam Propagation in the Atmosphere, *J. W. Strobehn, Ed.: Springer-Verlag*, 1978.

[103] C. B. Naila, A. Bekkali, K. Wakamori, and M. Matsumoto, "Performance analysis of CDMA-based wireless services transmission over a turbulent RF-on-FSO channel," *Journal of Optical Communications and Networking IEEE/OSA*, vol. 3, pp. 475-486, 2011.

[104] D. Chen, X. Z. Ke, and Q. Sun, "Outage probability and average capacity research on wireless optical communication over turbulence channel,"*10th International Conference on Electronic Measurement & Instruments (ICEMI)*, vol. 1, pp. 19-23, 2011.

[105] J. W. Goodman, Statistical optics. New York: *Wiley-Interscience*, January 18, 1985.

[106] A. Kolmogorov, "Turbulence," in Classic Papers on Statistical Theory, *S. K. Friedlander and L. Topper, Eds*. New York: *Wiley-Interscince*, 1961.

[107] A. Kolmogorov, "The local structure of turbulence in incompressible viscous fluid for very large Reynold numbers," *Proceeding of Royal Society of London Series A-Mathematical and Physical*, vol. 434, pp. 9-13, 1991.

[108] V. I. Tatarski, Wave propagation in a turbulent medium. (Translated by R.A. Silver-man) *New York: McGraw-Hill* 1961.

[109] A. Ishimaru, "The beam wave case and remote sensing," *in Topics in Applied Physics: Laser Beam Propagation in the Atmosphere*. vol. 25 p.134 New York: Springer-Verlag, 1978.

[110] N. D. Chatzidiamantis, H. G. Sandalidis, G. K. Karagiannidis, and M. Mathaiou, "Inverse gaussian modeling of turbulence-induced fading in freespace optical systems,"*IEEE/ OSA Journal of Lightwave Technology*, vol. 29, pp. 1590-1596, 2011.

[111] H. Hodara, "Laser wave propagation through the atmosphere," *Proceedings of the IEEE*, vol. 54, pp. 368-375, March 1966.

[112] J. W. Strohbehn, "Line-of-sight wave propagation through the turbulent atmosphere,"*Proceedings of the IEEE*, vol. 56, pp. 1301-1318, Aug. 1968.

[113] H. D.H., "Depolarisation of laser beam at 6328 Angstrom due to

atmospheric transmission," *Applied Optics*, vol. 8, p. 367, Feb. 1969.

[114] H. G. Sandalidis, "Coded free-space optical links over strong turbulence and misalignment fading channels," *IEEE Transactions on Communications*, vol. 59, pp. 669 674, 2011.

[115] E. S-Nasab, M.Uysal, "Generalized performance analysis of mixed RF/FSO system," *optical wireless communication 3rd international workshop*, pp.16-20, 2014.

[116] D. M. Forin, G. Incerti, G. T. Beleffi, A. Teixeira, L. Costa, P. D. B. Andrè, B. Geiger, E. Leitgeb, and F. Nadeem, "Free space optical technologies," *Telecommunication book' IN-TECH*, 2010.

[117] Z. Ghassemlooy, W. Popoola, and E. Leitgeb, "free-space optical communication using subarrier modulation in gamma-gamma atmospheric turbulence," in Transparent Optical Networks, *ICTON. 9th International Conference on*, pp. 156-160, 2007.

[118] M. A. Al-Habash, L. C. Andrews, and R. L. Phillips, "Mathematical model for the Irradiance probability density function of a laser beam propagating through turbulent media," *Optical Engineering*, vol. 40, pp. 1554-1562, 2001.

[119] S. G. Wilson, M. Brandt-Pearce, Q. Cao, and J. H. Leveque, "Free-space optical MIMO transmission with Q-ary PPM," *IEEE Transactions on Communications*, vol. 53, pp. 1402-1412, 2005.

[120] Hector E. Nistazakis, A.N. Stassinakis, Sinan Sinanovic, Wasiu O. Popoola, George S.Tombras," Performance of quadrature amplitude modulation orthogonal frequency division multiplexing-based free space optical links with non-linear clipping effect over gamma– gamma modelled turbulence channels", *in IET Optoelectronics*, pp. 1-6, 2015.

[121] J. H. Churnside and S. F. Clifford, "Log-normal Rician probability density function of optical scintillations in the turbulent atmosphere," *Journal of Optical Society of America*, vol. 4, pp. 1923-1930, 1987.

[122] L. C. Andrews and R. L. Phillips, "I-K distribution as a universal propagation model of laser beams in atmospheric turbulence," *Journal of*

optical society of America A, vol. 2, p. 160, 1985.

[123] G. Parry and P. Pusey, "K-distributions in atmospheric propagation of laser light," *Journal of the Optical Society of America*, vol. 69, pp. 796-798, 1979.

[124] Z. Ghassemlooy, W. Popoola, S. Rajbhandari, "Optical Wireless Communications: System and Channel Modelling with Matlab®", *CRC Press*, pp. 140-143, 2012.

[125] H. Zhang; H. Li; X. Dongya; C.Chaom, "Performance Analysis of Different Modulation Techniques for Free-Space Optical Communication System", *in Telkomnika*, Vol. 13 Issue 3, pp. 880-888, 2015

[126] M.R. Alam, S. Faruque, "Comparison of different modulation techniques for free space laser communication", *IEEE International Conference on Electro/Information Technology (EIT) on*, pp.637 – 640, 2015.

[127] K. P. Peppas and P. T. Mathiopoulos," Free-Space Optical Communication with Spatial Modulation and Coherent Detection Over H-K Atmospheric Turbulence Channels," *Journal of Lightwave Technology*, Vol. 33, pp. 4221-4232, 2015.

[128] S. Randel, F. Breyer, S. C. J. Lee, and J. W. Walewski, "Advanced modulation schemes for short range optical communications," *Selected Topics in Quantum Electronics, IEEE Journal of*, vol. 16, pp. 1280-1289, 2010.

[129] David J. Geisler; Curt M. Schieler ; Timothy M. Yarnall ; Mark L. Stevens; Bryan S. Robinson ; Scott A. Hamilton, "Demonstration of a variable data-rate free-space optical communication architecture using efficient coherent techniques", *Opt. Eng.* 55(11), Vol. 24, pp. 111605, 2016.

[130] A. K. M. Nazrul Islam ; S. P. Majumder, "Impact of timing jitter on the BER performance of an M-PPM free space optical link in presence of atmospheric turbulence"; *Electrical Engineering and Information Communication Technology (ICEEICT), International Conference on*, pp. 1-4, 2015.

[131] M. A. Islam, A. Bari and C. B. Barua, "Free-space optical communication

with m-ary pulse position modulation under strong turbulence with different type of receivers," *2nd International Conference on Electrical Information and Communication Technologies (EICT)*, pp. 259 - 262, 2015.

[132] S. Rajbhandari, "Application of wavelets and artificial neural network for indoor optical wireless communication systems, *Doctor of Philosophy, Northumbria University, UK*, 2009.

[133] L. W. Couch, H. Shao, X. Li, and L. Liu, Digital and analog communication systems: *Macmillan*, 1993.

[134] G. W. Marsh and J. M. Kahn, "Channel reuse strategies for indoor infrared wireless communications," *Communications, IEEE Transactions on*, vol. 45, pp. 1280-1290, 1997.

[135] M. D. Audeh and J. M. Kahn, "Performance evaluation of L-pulse position modulation on non-directed indoor infrared channels," *in Communications, SUPERCOMM/ICC'94, Conference Record, 'Serving Humanity through Communications. 'IEEE International Conference on*, pp. 660-664, 1994.

[136] R. Dettmer, "A ray of light, free space optical transmission," *IEE Review*, vol. 47, pp.32-33, 2001.

[137] T. O'Farrell and M. Kiatweerasakul, "Performance of a spread spectrum infrared transmission system under ambient light interference," in The Ninth IEEE International Symposium on Personal, *Indoor and Mobile Radio Communications*, pp. 703-707, 1998.

[138] Y. Cheng and S-H Hwang," Subcarrier Intensity Modulation/Spatial Modulation for Optical Wireless Communications", *IEICE TRANSACTIONS on Communications* Vol.E97-B No.5 pp.1044-1049, 2014.

[139] Md. Z. Hassan, Md. J. Hossain, J. Cheng and Victor C. M. Leung" Sub-carrier Intensity Modulated Optical Wireless Communications: A Survey from Communication Theory Perspective", *ZTE Communications*, Vol.14 No.2, pp.1-12, April 2016.

[140] M. Jazayerifar, I. Sackey, R. Elschner, F. Da Ros, T. Richter, C. Meuer, C. Peucheret, C. Schubert and K. Petermann, " Perspectives of long-haul WDM transmission systems based on phase-insensitive fiber-optic parametric amplifiers," *IEEE Summer Topicals Meeting Series (SUM)*, pp. 211 - 212 , 2015.

[141] M. A. Juang and M. B. Pursley, "Subcarrier Ordering for OFDM Packet Radio Systems," *IEEE Communications Letters*, Volume: 19, pp. 1434-1437, 2015.

[142] R. You and J. M. Kahn, "Average power reduction techniques for multiple Subcarrier Intensity modulated optical signals," *IEEE Transaction on Communications*, vol. 49, pp. 2164-2171, 2001.

[143] S.Zhang, X.You, Z.Hua and Y.Wang, "BPSK-SIM free-space optical communication with reference continuous wave light," *13th International Conference on Optical Communications and Networks (ICOCN)*, pp.1-3, 2014.

[144] X. Song and J. Cheng, "Subcarrier intensity modulated optical communications in strong atmospheric turbulence," *in Communications in China Workshops (ICCC), 1st IEEE International Conference on*, pp. 26-30, 2012.

[145] R. K.Giri and B. Patnaik, "A Comparative Analysis of Different Modulation Techniques Based on Subcarrier Intensity Modulation in Free Space Optics Using Log-Normal Turbulence Model," *International Conference on Information Technology (ICIT)*, pp.12-16, 2016.

[146] Douglas L. Jones, Swaroop Appadwedula, Matthew Berry, Mark Haun, Jake Janovetz, Michael Kramer, Dima Moussa, Daniel Sachs, Brian Wade, "Digital Transmitter: Introduction to Quadrature Phase-Shift Keying", *in Connexions module:* m10042, Version 2.19: US/Central, pp. 1-4, Feb 25, 2004.

[147] L.J. Ippolito,"Satellite Communications Signal Processing,"first edition, *Wiley Telecom eBook* Chapters, pp. 464, 2017.

[148] Esdras Anzuola. "Atmospheric compensation experiments on free-space

optical Coherent commnication systems", *Doctor of Philosophy, Polytechnical University of Barcelona*, 2015.

[149] M. S. Khan, S. S. Muhammad, M. S. Awan, V. Kvicera, M. Grabner, and E. Leitgeb, "Further results on fog modeling for terrestrial free-space optical links," *Optical Engineering*, vol. 51, pp. 031207-1, 2012.

[150] F. S. Marzano, P. Nocito, S. Mori, F. Frezza, P. Lucantoni, M. Ferrara, E. Restuccia, and G. M. T. Beleffi, "Characterization of hydrometeor scattering effects and experimental measurements using near-infrared free-space urban links," *in Antennas and Propagation (EUCAP) 6th European Conference on*, pp. 330-334, 2012.

[151] M. S. Khan, E. Leitgeb, R. Nebuloni, C. Capsoni, M. Grabner, and V. Kvicera, "Effects of PSA on free-space optical links," *in Antennas and Propagation (EUCAP), 6th European Conference on*, pp. 1244-1247, 2012.

[152] A. R. Raja, Q. J. Kagalwala, T. Landolsi, and M. El-Tarhuni, "Free-space optics channel characterization under uae weather conditions," in Signal Processing and Communications, *ICSPC 2007. IEEE International Conference on*, pp. 856-859, 2007

[153] W. O. Popoola, Z. Ghassemlooy, H. Haas, E. Leitgeb, and V. Ahmadi, "Error performance of terrestrial free space optical links with subcarrier time diversity," *Communications, IET*, vol. 6, pp. 499-506, 2012.

[154] D. K. Borah and D. G. Voelz, "Pointing error effects on free-space optical communication links in the presence of atmospheric turbulence," *Lightwave Technology, Journal of*, vol. 27, pp. 3965-3973, 2009.

[155] V. Brazda, O. Fiser, and J. Svoboda, "FSO and radio link attenuation: meteorological models verified by experiment," *Proceedings of the SPIE*, vol. 8162 pp. 81620N, 2011.

[156] A. Jurado-Navas and A. Garcia-Zambrana, "Efficient lognormal channel model for turbulent FSO communications," *Electronics Letters*, vol. 43, pp. 178-179, 2007.

[157] M. A. Khalighi, N. Schwartz, N. Aitamer, and S. Bourennane, "Fading

reduction by aperture averaging and spatial diversity in optical wireless systems," *IEEE/OSA Journal of Optical Communications and Networking*, vol. 1, pp. 580-593, 2009.

[158] A.Gatri, Z. Ghassemlooy, A.Valenzuela, O.Strobel, R.Rejeb and E.Leitgeb, "Experimental Study of the Performance for BPSK Subcarrier Intensity Modulation Free Space Optics Communications in a Laboratory Controlled Turbulence Channel," Network and Optical Communications (NOC), *18th European Conference on and Optical Cabling and Infrastructure (OC&i), 8th Conference on*, pp. 287-291, Graz, Austria, 2013.

[159] R. Pernice, A. Andò, M. Cardinale; L. Curcio; S. Stivala; A. Parisi; A.C. Bussacca; Z. Ghassemlooy; J.Perez, "Indoor free space optics link under the weak turbulence regime: measurements and model validation," *IET Communications*, Volume: 9, Issue: 1, pp.62-70, 2015.

[160] M. S. Awan, L. C. Horwath, S. S. Muhammad, E. Leitgeb, F. Nadeem, and M. S. Khan, "Characterization of fog and snow attenuations for free-space optical propagation," *Journal of Communications*, vol. 4, pp. 533-545, 2009.

[161] M. S. Awan, E. Leitgeb, M. Loeschnig, F. Nadeem, and C. Capsoni, "Spatial and time variability of fog attenuations for optical wireless links in the troposphere," *in IEEE 70th Vehicular Technology Conference Fall (VTC)*, pp. 1-5, 2009.

[162] M. Grabner and V. Kvicera, "On the relation between atmospheric visibility and optical wave attenuation," *in Mobile and Wireless Communications Summit, 16th IS*T, pp. 1-5, 2007.

[163] Uysal, M., Capsoni, C., Ghassemlooy, Z., Boucouvalas, A. C., and Udvary E. G. (Eds.): Optical Wireless Communications–An Emerging Technology, *Springer*, ISBN: 978-3-319-30200-3, 2016.

[164] X. Wu, P. Liu, and M. Matsumoto, "A study on atmospheric turbulence effects in full optical free-space communication systems," *6th International Conference on Wireless Communications Networking and*

Mobile Computing 2010 (*WiCOM*), pp. 1-5, 2010.

[165] S. Zvanovec, J. Perez, Z. Ghassemlooy, S. Rajbhandari, and J. Libich, "Route diversity analyses for free-space optical wireless links within turbulent scenarios," *Optics Express*, vol. 21, pp. 7641-7650, 2013.

[166] M. Ijaz, O. Adebanjo, S. Ansari, Z. Ghassemlooy, S. Rajbhandari, H. Le Minh, A. Gholami, and E. Leitgeb, "Experimental Investigation of the Performance of OOK-NRZ and RZ Modulation Techniques under Controlled Turbulence Channel in FSO Systems," *IEEE Trans*, 2010.

[167] H. Burris, A. Reed, N. Namazi, W. Scharpf, M. Vicheck, and M. Stell, "Adaptive thresholding for free-space optical communication receivers with multiplicative noise," *in Aerospace Conference Proceedings, IEEE*, pp. 3-1473-3-1480 vol. 3, 2002.

[168] Rabinovich WS, Moore CI, Mahon R, Goetz PG, Burris HR, Ferraro MS, Murphy JL, Thomas LM, Gilbreath GC, Vilcheck M, Suite MR, "Free-space optical communications research and demonstrations at the U.S. Naval Research Laboratory," *Journal of Applied Optics* vol. 54(31), pp. 189-200, 2015.

[169] M. Faridzadeh; A. Gholami; Z. Ghassemlooy; A. Gatri, "BPSK-SIM-PPM modulation for free space optical communications," *Telecommunications (IST), 7th International Symposium on*, pp. 794 - 798, 2014.

[170] Z. Ghassemlooy, W.O. Popoola, S. Rajbhandari, M. Amiri, S. Hashemi,: "A synopsis of modulation techniques for wireless infrared communication", Invited paper, *IEEE-International Conference on Transparent Optical Networks, Mediterranean Winter (ICTON-MW)*, Dec, 6-8, 2007-Sousse, Tunisia, pp. 1-6.

[171] E. M. Abdullah; F. H. Alouini, M.Slim, "Outdoor FSO Communications Under Fog: Attenuation Modelling and Performance Evaluation", *IEEE Photonics Journal* vol.8(4):1-1, Volume 8, Number 4, pp. 1-22, 2016.

[172] J. Vitasek, J. Látal, V. Vašinek, S. Hejduk, A. Liner, M. Papes, P. Koudelka, and A. Ganiyev, "The fog influence on bit error ratio,"

Proceedings Volume 8697, 18th Czech-Polish-Slovak Optical Conference on Wave and Quantum Aspects of Contemporary Optics pp. 86970L, 2012.

[173] M. Ijaz, Z. Ghassemlooy, H. Le Minh, S. Rajbhandari, and J. Perez, "Analysis of fog and smoke attenuation in a free space optical communication link under controlled laboratory conditions," *in Optical Wireless Communications (IWOW), International Workshop on*, pp. 1-3, 2012.

[174] S. Mori, F. Marzano, F. Frezza, G. Beleffi, V. Carrozzo, A. Busacca, and A. Ando, "Model analysis of hydrometeor scattering effects on free space near infrared links," *in Optical Wireless Communications (IWOW),International Workshop on*, pp. 1-3, 2012.

[175] N. A. M. Nor, I. M. Rafiqul, W. Al-Khateeb, and S. A. Zabidi, "Environmental effects on free space earth-to-satellite optical link based on measurement data in Malaysia, "*in Computer and Communication Engineering (ICCCE), International Conference on*, pp. 694-699, 2012.

[176] F. Nadeem, M. Khan, and E. Leitgeb, "Optical wireless link availability estimation through Monte Carlo simulation," *in Telecommunications (ConTEL), Proceedings of the 11th International Conference on*, pp. 345-350, 2011.

[177] M. Awan, E. Leitgeb, F. Nadeem, M. Khan, and C. Capsoni, "A new method of predicting continental fog attenuations for terrestrial optical wireless link," in Next Generation Mobile Applications, *Services and Technologies, NGMAST. Third International Conference on*, pp. 245-250, 2009.

[178] M. Grabner and V. Kvicera, "The wavelength dependent model of extinction in fog and haze for free space optical communication," *Journal of Optics Express*, vol. 19, pp. 3379-3386, 2012.

[179] F. Nadeem, T. Javornik, E. Leitgeb, V. Kvicera,G. Kandus, „Continental Fog Attenuation Empirical Relationship from Measured Visibility Data", *in Journal of Radio Engineering,* Vol. 19 Issue 4, pp. 596-600, 2010.

[180] M. Ijaz, Z. Ghassemlooy,H., Le Minh, and S. Rajbhandari, "Bit error rate measurement of free space optical communication links under laboratory-controlled fog conditions", *NOC 16th European Conference on*, pp. 52-55, 2011.

[181] W. Popoola, Z. Ghassemlooy, M. S. Awan, and E. Leitgeb, "Atmospheric channel effects on terrestrial frees pace optical communication links," *in Proceedings of 3rd International Conference on Electronics, Computers and Artificial Intelligence*, Pitesti, Romania, , pp. 17-23, 2009.

[182] M. Al-Naboulsi, F. De Fornel, E. Leitgeb, S. Sheikh Muhammad et al., "Measured and predicted light attenuation in dense coastal upslope fog at 650, 850 and 950 nm for FSO applications," *Journal of Optical Engineering*, vol. 47, no. 3, pp.036001-1–036001-14, 2008.

[183] A. Arnulf et al., "Transmission by Haze and Fog in the Spectral Region 0.35 to 10 Microns," *Journal of Optical Society of America*, vol. 47, No. 12, pp. 491-498, June 2001.

[184] E. Wainright, H. H. Refai, H. Hazem, and J. J. Sluss, Jr., "Wavelength diversity in free-space optics to alleviate fog effects," *Proc. of SPIE*, vol. 5712, pp. 110-118, Ap. 2005.

[185] M. Achour, "Free-space optics wavelength selection: 10u versus short wavelengths," *Proc. of SPIE*, vol. 5160, pp. 1–15, 2003.

[186] W. K. Pratt, Laser Communication Systems, *J. Wiley & Sons*, New York, 1969.

[187] H. Weichel, Laser Beam Propagation in the Atmosphere, *SPIE*, Bellingham WA, 1990.

[188] G. Rousset, "Wave-front Sensors," in Adaptive Optics in Astronomy, F. Rouddier, Eds., Cambridge, UK: *Cambridge University Press*, pp. 99-130, 1999.

[189] L. Ming and M. Cvijetic, "Coherent free space optics communications over the maritime atmosphere with use of adaptive optics for beam wavefront correction," *Applied Optics*, Vol.54, Issue. 6, pp. 1453- 1462, 2015.

[190] H. Manor and S. Arnon, "Performance of an optical wireless communication system as a function of wave length," *J. Applied Optics*, vol. 42, no. 21, pp. 287-289, Dec. 2002.

[191] Xian Liu, "Performance of the wireless optical communication system with variable wave length and Bessel pointing loss factor," *proceeding of Wireless Communications and Networking Conference* (*WCNC*), Las Vegas, pp. 797–802, 2008.

[192] M. S. Awan, Marzuki, E. Leitgeb, F. Nadeem, M. S. Khan, and C. Capsoni, "Weather effects impact on the optical pulse propagation in free space," *Proceeding of the 69th Vehicular Technology Conference* (*VTC*), Barcelona, pp. 1–5, 2009.

[193] F. Xu, M.-A. Khalighi, and S. Bourennane, "Impact of different noise sources on the performance of PIN- and APD-based FSO receivers," *Telecommunications* (*ConTEL*), *Proceedings of the 11th International Conference*, Graz, pp. 211-218, 2011.

[194] M. Uysal and J. Li, "BER performance of coded free-space optical links over strong turbulence channels," *Proceeding of the 59th Vehicular Technology Conference (VTC)*, Milan, pp. 352–356, 2004.

[195] M. Faridzadeh; A. Gholami; Z. Ghassemlooy; S. Rajbhandari, "Hybrid PPM-BPSK subcarrier intensity modulation for free space optical communications," *Journal of the Optical Society of America. A, Optics, Image Science, and Vision*, vol. 29 (8), pp. 1680-1685, 2012.

[196] T. Ohtsuki, "Multiple-subcarrier modulation in optical wireless communications, "*IEEE Communication Magazine*, vol. 41, pp. 74-79, 2003.

[197] S. V. Kartalopoulos, "Free space optical networks for ultrabroad band services, "*J. Wiley & Sons*, New Jersey, pp. 35-36, 2011.

[198] M. Jonasz and G. Fournier, "Light scattering by particles in water: theoretical and experimental foundations," *Elsevier Inc*, Oxford, UK, 2007.

[199] A.Touati, A. Abdaoui, F. Touati, A. Bouallegue, "On the Effects of

Combined Atmospheric Fading and Misalignment on the Hybrid FSO/RF Transmission," *in Journal of Optical Communications and Networking* vol. 8(10), pp. 715-725, 2016.

[200] A. Eslami, S. Vangala, H. Pishro-Nik," Hybrid Channel Codes for Efficient FSO/RF Communication Systems," *in IEEE Transactions on Communications*, pp. 2926-2938, 2010.

[201] T. T.Nguyen, L.Lampe," Channel Coding Diversity with Mismatched Decoding Metrics," *in IEEE Communications Letters*, pp. 916-918, 2011.

[202] K. Kaur,R. Miglani,G. S.Gaba," Communication theory review perspective on channel modeling, modulation and mitigation techniques in free space optical communication," *in International Journal of Control Theory and Applications* vol. 09(11), pp. 4969-4978, 2016.

[203] K.O. Odeyemi,P.A. Owolawi,V.M. Srivastava," Performance analysis of free space optical system with spatial modulation and diversity combiners over the Gamma Gamma atmospheric turbulence," *in Optics Communications Journal*, pp. 205-211, 2017.

[204] J.Park, G. Park, B. Roh, G. Yoon, "Performance Analysis of Asymmetric RF/FSO Dual-hop Relaying Systems for UAV Applications," *IEEE Military Communications Conference*, pp. 1651-1656, 2013.

[205] J. Bohata,P. Pesek,S.Zvanovec,Z.Ghassemlooy," Extended measurement tests of dual polarization radio over fiber and radio over FSO fronthaul in LTE C-RAN architecture," *IEEE 12th International Conference on Wireless and Mobile Computing, Networking and Communications (WiMob)*, pp. 476-481, 2016.

[206] http://uk.ni.com/, access date: 11/02/2017

[207] Collet, E. & Alferness, R,"Depolarization of a laser beam in a turbulent medium,"*Journal of the Optical Society of America*. vol. 62, No. 4, pp. 529-533, 1972.

[208] Strohbehn, J.W., y Clifford, S,"Polarization and angle-of arrival fluctuations for a plane wave propagated through a turbulent medium", *IEEE Transactions on Antennas and Propagation*, vol. 15, No. 3,pp. 416-

421, 1967.

[209] Wheelon, A. D. Electromagnetic scintillation Vol, I. *Geometrical optics*. Cambridge University Press. Cambridge, 2001.

[210] S. S. Muhammad, B. Flecker, E. Leitgeb, and M. Gebhart, "Characterization of fog attenuation in terrestrial free space links," *Journal of Optical Engineering*, vol. 46, pp. 066001-066006, 2007.

[211] Primmerman, C., A., "Compensation of atmospheric optical distortion using a synthetic beacon", *Nature*, 353(6340), pp. 141-143, 1991.

[212] Barchers, J., D., Fried D., L., "Optimal control of laser beams for propagation through a turbulent medium", *JOSA A*, 19, pp. 1779-1793, 2002.

[213] Schwartz, N., et al., "Mitigation of atmospheric effects by adaptive optics for free-space optical communications", *proc. of SPIE*, pp. 72000J-72000J-11, 2009.

[214] A. Montmerle Bonnefois, R. Bi´erent, M. Raybaut, A. Godard, S. Derelle, A. Dur´ecu, V. Michau, M. Lefebvre, N. V´edrenne, and M.-T. Velluet, "SCALPEL: a long range free-space optical communication system with adaptive optics in the MIR bandwidth," *in Society of Photo-Optical Instrumentation Engineers (SPIE) Conference Series*, vol. 7828, pp. 78280L 2010.

[215] http://www.produktinfo.conrad.com/datenblaetter/625000649999/646188-an-01-de-FUNKWETTERSTATION_WS1600.pdf, last visited October 2017.

[216] https://www.e-wetter.eu/vpzubehor/weatherlink-datenlogger--software.php, last visited October 2017.

[217] M. Grabner and V. Kvicera, "Fog attenuation dependence on atmospheric visibility at two wavelengths for FSO link planning," *in Antennas and Propagation Conference (LAPC)*, Loughborough, pp. 193-196, 2010.

[218] K. W. Fischer, M. R. Witiw, and E. Eisenberg, "Optical attenuation in fog at a wavelength of 1.55 micrometers," *Atmospheric Research*, vol. 87, pp.

252- 258, 2008.

[219] H. Kaushal, G. Kaddoum, "Optical Communication in Space: Challenges and Mitigation Techniques," *IEEE Communications Surveys & Tutorials*, vol: 19, Issue: 1, pp. 57 - 96, August 2016.

[220] B. Reiffen and H. Sherman, "An optimum demodulator for Poisson processes: Photon source detectors," *Proc. IEEE*, vol. 51, pp. 1316–1320, 1963.

[221] S.Rajbhandari, Z. Ghassemlooy, P. A.Haigh, T. Kanesan, and Tang., "Experimental Error Performance of Modulation Schemes Under a Controlled Laboratory Turbulence FSO Channel", *Journal of Lightwave Technology*, Vol. 33, pp. 244-250, Jan 2015.

[222] H. Kaushal, V.K.Jain, and S. Ka, " Effect of atmospheric turbulence on acquisition time of ground to deep space optical communication system, *International Journal of Electrical and Computer Engineering*, vol. 4, pp. 730-734, 2009.

[223] Thorlabs Inc, (USA), www.thorlabs.com/newgrouppage9.cfm?object group_id=3257&pn=PDA36A, last visited October 2017.